Control System Migrations: A Practical Project Management Handbook

Control System Migrations: A Practical Project Management Handbook

DANIEL ROESSLER

MOMENTUM PRESS, LLC, NEW YORK

Control System Migrations: A Practical Project Management Handbook

First published by Momentum Press®, LLC
222 East 46th Street, New York, NY 10017
www.momentumpress.net

ISBN-13: 978-1-60650-443-7 (hard cover, case bound)
ISBN-10: 1-60650-443-6 (hard cover, case bound)
ISBN-13: 978-1-60650-445-1 (e-book)
ISBN-10: 1-60650-445-2 (e-book)

DOI: 10.5643/9781606504452

Cover Design by Jonathan Pennell
Interior design by Exeter Premedia Services Private Ltd., Chennai, India

10 9 8 7 6 5 4 3 2 1

Printed in the United States of America.

To my Mom and Dad who established the strong foundation of values on which I rely every day. You have given me the gifts of independent thinking, perseverance, and a strong work ethic. Thank you for your support, love, and encouragement throughout my life.

Contents

List of Tables and Figures

Acknowledgments

The author acknowledges the following individuals, who have enriched this book by sharing their knowledge and encouragement through reviews, suggestions, and/or informative discussions:

Dr. Vassilios Tzouanas—For your review, insightful comments, and feedback. You are a respected colleague and valued friend.

Nigel James—For your enthusiasm, excitement, and encouragement about this project. Thank you for discussing your ideas with me, for writing a thoughtful foreword, and for your treasured friendship.

Billy Adney—For reviewing the book, providing valuable feedback, and willingly sharing your vast process control knowledge and experience with me through the years. It is a pleasure to call you my friend.

Joel Stein—For your eagerness to work with me on this project and contributing your extensive technical publishing expertise.

Cindy Durand—For your review of my drafts and your swift responsiveness.

Millicent Treloar—For sharing your feedback and offering guidance to improve this book.

Acknowledgments

Foreword

The release of this book provides the automation industry a comprehensive guide through the complexities of control system migration projects. It should be a required reference for all project and program managers, engineers, and other resources involved in migrations. The book is insightful, methodical, well-written, and based on actual experiences with meaningful anecdotes and realistic examples. Whether you are part of an end user, engineering company, system integrator, or vendor organization, this book will be a valuable tool.

I have spent my entire career of almost 30 years in the controls and automation arena working directly for major refineries as well as managing system integration and automation service companies. I have learned that control system migration projects are among the most complex that most control engineers face in their career. From the blended resource teams required, to addressing customization and interface details, the challenges are numerous. My personal involvement in countless migration projects has taught me that each one is unique, but there are certain methodologies and best practices that are applicable across the board.

I worked as Dan's manager at a system integration company for a period of time and have known him for over 15 years. He is the perfect author for this book on managing control system migration projects. I can vividly remember a project in the pulp and paper industry where developing and using these concepts helped save the project after some early issues. I also remember a refinery operations manager once stating of Dan's work, "This is the best control system migration FEL I have ever read."

I know firsthand that this book is the fruit of numerous nights in the control room. It evolved over a number of years while Dan was experiencing the challenges of managing control system migration projects. The wealth of experience shared in this book is formed as much from the failures and defeats as from the successes. I am proud to see his journey result in a resource that will simplify the migration experience for others.

We are entering another golden age of engineering, innovation, and project work. Our work force is aging, and the speed to market of new technology is at its fastest pace in history. How do we fill the knowledge gap and hand down to our younger brethren the value of our experiences? That is exactly what this book does. I believe reading it will establish a firm foundation for control system migration success. I plan to make this book required reading for all my staff.

Theodore Roosevelt once stated, "The credit belongs to the man who is actually in the arena." Dan is one of those men who dares greatly and has spent himself on a worthy cause.

Nigel James
President-Burrow Global Automation

Preface

In 1992, I had just earned my degree in electrical engineering and was beginning full-time employment at a chemical (polymer) plant that I had interned at the previous two summers. The plant was in the midst of a control system migration and I was given an opportunity to help configure the new system under the watchful guidance of a senior controls engineer. It was my first migration project and a great learning experience. That project taught me a lot about the process of converting points, graphics, and control logic from an older system to a newer system including some of the unique challenges that migration projects present.

The world of process automation has changed significantly in the 20 years since my first migration project. The most visible changes are in technology. We have shifted from proprietary, largely independent control systems, somewhat isolated from other plant systems to PC-based solutions using standard operating systems. Today's control systems now have more open connectivity and are commonly integrated with numerous third-party systems and applications. They are also much easier to configure with extensive standard features and functions, minimizing the need for customization.

The way that control systems are used has also evolved. As a result, end users now expect much more from their control systems than they once did in most areas, such as reliability, flexibility, speed, and functionality. And, importantly, the process of making control systems purchasing decisions has changed. Now that the differences in control systems offered by various vendors are more subtle, the system selection process requires end-users to more clearly understand and prioritize their unique control system needs and decision criteria.

As legacy control systems move toward the end of their lifecycles and in some cases reach obsolescence, migrations to newer technologies are becoming more prevalent. Even those companies which might prefer to continue extending the life of their existing control system are often compelled to migrate because of system limitations related to critical issues like cyber security. As a result, companies across process industries are currently making large financial investments to replace older control system infrastructure. This trend is expected to continue for the foreseeable future. The pace of technology advancements means that even some newer control systems will reach the end of their lifecycle faster and require replacement or upgrade sooner.

There are many variations on the definition of control system migrations largely driven by the motivation of the individual tasked with defining it. For the purposes of this book, I have included what some refer to as modernizations because I believe the same considerations and project management challenges apply whether you are converting an older control system to a different vendor's system or upgrading to your existing vendor's latest solution. Control system migrations update older systems to newer technology by substantially changing both the hardware and software of a control system. Minor software version upgrades are not considered migrations.

Note that the term control system generally refers to the combination of hardware and software technology that provides the ability to operate and control field devices from a centralized location. This includes Distributed Control Systems (DCSs) as well as Programmable Logic Controllers (PLCs), and Human Machine Interface (HMI) systems. There are also other commonly used control systems such as Safety Instrumented Systems (SISs) and Supervisory Control and Data Acquisition (SCADA) systems. For the purposes of this book, I group all of these systems and refer to them using the common term of control systems. While there are some differences in the detailed functionality, purposes, applications, and architectures of these systems, the process of migrating them to newer technologies requires the same basic process which is the focus of this book.

Control system migration projects have many obstacles that are not always obvious. This handbook discusses some of these unique challenges and recommends various tools and approaches for handling them. Control system migrations require you to understand and document all aspects of the configuration within the old system as well as develop a transition plan to the new system, often times in phases, while minimizing impact to manufacturing production. The success of your migration project will be greatly influenced by the choices that you make about when and how you move from the planning phase to the execution phase and finally to the operational phase of your project.

Fundamental project management principles apply to control system migration projects just as they apply to other disciplines, such as civil or mechanical projects. In addition, good project management requires understanding subtleties like:

- When a scope is the right granularity?
- When a schedule is realistic and accounts for hidden "gotchas"?
- How to smoothly transition through the various phases of a project?

The project team is another key component of any project's success. Team members must understand the overall project and be clear on their specific responsibilities. This handbook is intended to educate all parties on key areas of migration projects and provide insight into how to best plan and execute the project. I hope that you will refer back to this handbook often as your preferred resource in guiding you to the successful completion of your control system migration project.

Effectively managing control system migration projects is about a process and successful projects are a result of many factors. There are many ways to approach a control system migration project and no two control system migration projects are identical. The chances of success are directly related to good planning and the approaches used to manage and execute the project. This is true of most projects and certainly applies to control system migration projects. The project manager must be familiar enough with the project to know what is reasonable and be able to proactively identify and manage high risk areas.

If you are new to control system migrations, this book should advance your understanding of the migration projects steps, help you identify key areas on which to focus time and effort, and provide you with tools to better plan and execute your project. If you are a control system migration veteran hopefully you will find many familiar concepts and approaches along with a few unique perspectives and valuable suggestions. It is my hope that this book delivers insight into control system migration projects to a broad audience of end users, system integrators, EPC companies, vendors, and control system engineering students.

Whether you are an operator helping design graphics as part of the control system migration team or the instrument and electrical (I&E) manager responsible for the field installation, this book will provide an overview of the steps and stages of the migration process, tips and suggestions for success, and a better appreciation of your role in the project. Managers in operations, maintenance, and engineering departments should also find this book helpful in better understanding the value and benefits of a control system migration. Improving the understanding of all parties involved in a control system migration project will help each understand how they influence the project's success. When the project is successful, the entire team wins.

My goal with this handbook is to capture migration best practices and ensure successful transitions from one control system to the next. The book is organized in a logical project workflow and begins by examining how to build an effective justification that will help initiate your control system migration project. We then cover the details of performing a comprehensive Front End Loading (FEL) study which forms the critical foundation of a successful control system migration project. Our next focus is on how to build complete and effective scope, schedule, and budget documents that are consistent relative to one another, enhancing your chances for a successful migration. There is also valuable information included to help guide users through a logical vendor selection process that reduces the emotional component of decision-making which is commonly a source of frustration on migration projects.

The book includes a chapter on how to plan and execute training which is often an overlooked aspect of migration projects essential to success. We review common migration project challenges related to graphics, third-party application integration, and numerous other high risk areas detailing specific issues and discussing ways to avoid them. We also examine the nuances of successfully planning, managing, and executing a system cutover which is generally the area of highest risk on a migration project.

Many processes are outlined, templates provided, and topics discussed specific to project management activities in this handbook. We cover key elements of project staffing and how to build an effective team. We also examine how to handle project monitoring, change order management, and project reporting throughout the project detailing how to make these project management responsibilities an extension of normal project activities reducing the amount of time and effort required. Finally, we address the project closeout process and how to transition from the project to an effective lifecycle management program for your new control system.

The content of this book includes methods, approaches, and tools that have worked for me on specific migration projects throughout my career. You will want to selectively use and adapt this information so that it works best for your particular project. It is my hope that by identifying and outlining key considerations of a control system migration project and giving context to these elements, you will be better prepared for migration success.

This handbook should help you, whether or not you are familiar with control systems, to approach the migration project in a methodical manner. This book contains boxed item features of some of my experiences in anecdotes and examples. I also outline some ways to establish common expectations among all parties early in the project process so that everyone is aligned and working toward the same goals.

Many of the tables in this book can be used as checklist by your project team. Project team members cannot know or remember everything so this is written as a handbook that will enable you to quickly reference specific areas when needed. Each chapter of the book begins with an overview of the topics and ends with a summary of key takeaways, similar to the following takeaways from this Preface:

Three Key Takeaways

- This handbook is intended for a broad audience of people with diverse roles either directly or indirectly involved in control system migration projects.
- Topics covered in this book are comprehensive and include everything from the migration project justifications to vendor selection processes to project management reporting best practices.
- The processes, tools, and tips shared in this book are a result of my experiences with migration projects for more than 20 years as an end user, system integrator, and control system vendor.

KEY WORDS

Control System Migration, DCS Migration, Control Room Modernization, SCADA Upgrade, Cutover, Front End Loading, DCS System, Control System Projects, Project Management Methodologies, Project Management Tools and Techniques, Project Management Process, Project Management Books, Project Management Best Practices, Project Management Resources, Control System, Control System Design, Control Systems Engineering, Distributed Control System, Process Control Systems, Control System Upgrades

About the Author

My early career was as a controls engineer and project manager for a chemical company. I worked in two plant locations over roughly nine years. Our control systems included a combination of DCS and PLC systems that were operated from multiple control rooms. During my time at the plants, I managed several migration related projects including DCS controller upgrades, a control room consolidation, and operator workstation migrations. Our team also had responsibilities for the configuration and maintenance of these control systems as well as third-party systems and applications interfaced with the DCS, such as the process historian and advanced process control solutions.

The next six years of my career were at a system integration company. It was during this time that I was exposed to migration projects that involved multiple industries, numerous control systems, and diverse scopes. My responsibilities included leading migration Front End Loading (FEL) studies and managing migration projects. There are noticeable differences among industries and systems, but I also learned that there are many more commonalities which contribute to control system migration project success. I was also responsible for proposal development and as a result understand the details of the bid process. This experience has helped me understand how to position bid documentation to get the most complete and comparable proposals from control system vendors and services providers.

Finally, I have spent a number of years working for process automation vendors. This includes several years at a major control system vendor as well as time with software vendors providing applications that integrate with control systems. I learned a tremendous amount about challenges that control system migration projects present to the vendors themselves. I also gained insight into the strategies and approaches that vendor's use in responding to bids. As a result, I can offer tips on avoiding common misunderstandings during the bid process and streamlining control system vendor selection.

I have learned through my diverse work experiences that all parties, whether the control system vendor, a system integrator, or the end user, want a successful control system migration. Unfortunately, not everyone has the same definition of success, hence the need for a handbook to help establish alignment among all team members. My experiences are unique to me and my perspectives are a direct result. After over 20 years involved in some way with control system migration related projects, I am confident that the strategies, approaches, and processes that I provide in the following pages are a roadmap to successfully managing your migration project.

1

Migration Project Justification

It is a testament to the automation vendors of the 1970s, 1980s, and 1990s that so many legacy control systems from these periods continue to operate industrial facilities today. However, many process control engineers will tell you this is both a blessing and a curse. Because these control systems continue to operate reasonably well, one of the biggest challenges controls engineers face is getting support for control system migrations. Short of failures by the control system that lead to significant downtime, the need for migration projects is often scrutinized and questioned.

When compared with capital expenditures on more tangible and easily understood return-on-investment (ROI) projects, control system migrations are frequently considered lower priorities, which often results in repeated delays to funding them. Understanding the ROI of a project like an equipment debottleneck is straightforward because additional throughput capacity is easily converted to dollars. The ROI on control system migration projects tends to be much less tangible and more difficult to convert to financial benefits that are easily agreed on by key decision-makers.

It is important that someone take ownership of building the case for justifying a control system migration while also establishing realistic expectations within the organization regarding the benefits of replacing the control system with newer technology. This person is often the controls engineer who may or may not be the project manager on the eventual control system migration project. For example, some organizations have a separate project group to manage all capital projects. The project manager, if someone other than the controls engineer, should be identified and involved in the justification process if at all possible. This is the individual who will ultimately be responsible for ensuring that the project delivers on the economics defined in the Authorization for Expenditure (AFE) process.

In this chapter, we examine how to build an effective justification so that your control system migration project will be appealing. We begin by discussing different funding request strategies as well as how to identify and involve key stakeholders that can help support justification efforts. Determining the appropriate timing for your control system migration is also examined. We review how to capture your control system migration project's ROI and develop supporting business cases and examples. Finally, detailed sections are included, which highlight common areas to consider in your justification process such as parts availability issues and limitations in integrating with third-party systems and applications.

DETERMINING YOUR APPROACH

One of the first steps in the justification process is determining what funding to initially request. There are two options, either request funding for the full project or request funding for a Front End Loading (FEL) study, to better define the scope, schedule, budget, and other project details. Requesting FEL funding is generally a better approach. It requires substantially less initial funds so is often easier to gain approval. Also, a good FEL study details the scope, identifies areas of risk and uncertainty, and generates a tighter estimate with reduced variability. When the subsequent full migration project funding is requested, there is typically much greater certainty in the estimated project cost increasing the confidence of decision-makers and approvers. Even if the migration project is not immediately approved, the FEL documentation is usually largely applicable when the project does move forward requiring minor revisions to account for any updates.

When requesting full funding without having completed a thorough FEL, the budget is much more at risk and typically not more detailed than ±25%. For some organizations, this is an acceptable approach. If this approach is used, be sure to include money for upfront FEL work as part of the overall scope and build this upfront engineering effort into the schedule. This step is crucial to help identify and resolve potential problem areas before project execution begins.

It is essential to include as many key stakeholders as possible in developing the business case for a control system migration project. Identifying parties with a vested interest early in the process and understanding their control system needs not only helps you build a compelling business case but also strengthens support for the project. For instance, visit with the maintenance manager, instrument and electrical (I&E) supervisor, and I&E technicians to understand what challenges they have in maintaining the system. Maybe the team is having trouble with common parts availability on input/output (I/O) cards, controllers, termination panels, or operator consoles? Maybe they have been purchasing updated field instrumentation with expanded diagnostic capability and can't take full advantage of these newer diagnostics in the older control system? The team will appreciate you seeking their input and will be more supportive of the project if they believe it will help alleviate some of their specific work challenges.

Operations, maintenance, and engineering are the common organizations represented in the utilization and care of most control systems. While not as obvious, there are other groups to include as well. Information Technology (IT) is often involved in getting data out of the control system for other applications. Increasingly, the separation of company IT and control system networks are blurred. While most controls engineers will argue that the process control network (PCN) is a separate entity, at the very least, critical information is transferred across networks daily to support business applications. Talking with the owners and users of various business applications that utilize control system data is also important. They may have difficulty getting the information they need from the control system, may need different formatting or granularity, or have other challenges that can also contribute to the business justification for migrating the control system.

Another factor in the justification process is identifying the proper timing to propose the migration. The timing of when a control system migration should occur is not an exact science. Control system viability is based on balancing a combination of factors such as reliability, total cost of ownership, and system performance to meet defined business objectives. Whenever a given element is out of balance, it can signal that it is time to evaluate whether a migration might be necessary.

Obviously, there are numerous factors in the migration timing consideration process. Each company must make the decision based on the available capital and relative prioritization with other projects. Some companies will take a pro-active approach and continually evaluate the long-term viability of their control systems. Other companies hold on to their existing control systems well past the optimal point and are hesitant to migrate until a specific issue, such as security vulnerability, system failures, etc. forces them to take action. It is important that the decision to migrate your control system is an educated one that is based on a full understanding of not only the financial ROI of a migration project but also those benefits that may be difficult to quantify. Every company should have a control system lifecycle plan that is reviewed and updated at a regular frequency, which will help them evaluate the lifecycle status of their system.

DEFINING ROI

So how do you calculate ROI for a migration project? There is no universally agreed upon answer. Companies use a variety of ways to calculate ROI, rate-of-return analysis, and other financial metrics to determine whether a proposed project meets defined payback thresholds. I won't tackle ROI calculation methodology details other than to note that it is seldom a single factor that should be used to justify a control system migration project. Instead a combination of additive factors is generally used to build a strong ROI case. There are of course projects that are approved based on factors other than ROI, such as safety and maintenance reliability projects. In these situations, many of the justification cases outlined below will still apply.

For many industrial facilities today, control system migrations are part of a larger vision. For instance, facility siting frequently identifies problematic control room locations. As a result, companies are building new control rooms in alternate locations within the facility and using this opportunity to upgrade their control system. Many companies have also reduced staffing and subsequently consolidate control rooms as part of streamlined operations. In these cases, control rooms are often re-designed and control systems are updated as part of the consolidation process as well.

When there are larger projects, such as these driving control system replacements or modernizations, it can reduce the challenges of the justification process but comes with other pitfalls. In these situations, the control system migration is not the focal point of the project and the scope as well as the budget can get minimized to balance other parts of the project. If scope or budget reductions to the control system migration occur and are substantial they can impact the long-term benefits of the migration. It is also not uncommon to see critical control related design strategies or control system selections being made for financial rather than technical reasons with prioritization given to how it impacts the overall project rather than what is ideal for long-term operations.

What are some effective justifications for standalone control system migration projects? The answer is complex because it is largely dependent on a given facilities specific situation. Some common considerations that contribute to control system migration project justifications are identified in Table 1.1 below.

Table 1.1. Common control system migration project ROI considerations

Issue	Result	ROI Considerations
• System failures	• Decreased reliability • Increased downtime	• Lost production • Unplanned outages • Product schedule and shipment disruptions
• Parts availability or obsolescence issues	• Extended outages (often unplanned) • Lost system functionality • Increased maintenance requirements	• Increased maintenance costs • Lost production • Impact to production quality
• Difficulty integrating with newer applications and systems	• Can't realize full potential of new applications • Key data not easily available to decision-makers	• Less optimized operational performance • Slower business decisions • Higher costs associated with project implementations and ongoing support
• Reduced availability of support services	• Difficulty troubleshooting maintenance issues • Extended schedules for projects requiring engineering	• Increased maintenance costs • Increased engineering costs • Delays in realizing benefits of projects involving control system configuration
• Operational inefficiency	• Inability to take advantage of current best practices • Operator mistakes contributing to product quality issues and downtime	• Reduced product quality • Increased downtime • Operator stress

As you begin to put together your justification, consider how many of the considerations in Table 1.1 are applicable to your control system migration project. The ROI for your specific project will be unique and may include numerous other issues related to your particular migration. We examine these common issues that drive control system migration projects in more detail throughout the remaining sections of this chapter.

SYSTEM FAILURES

The most straightforward justification for a control system migration is when the system is failing. When your plant shuts down multiple times in a year due to control system failures, the justification process becomes immediately easier. However, no controls engineer wants the frustrations or visibility that comes with these system failures. If this is the situation at your company, then it is possible that you are well past the best time to migrate your control system. Ideally, your migration project takes place prior to reaching a stage in the lifecycle where an unacceptable failure rate is occurring. As indicated in Figure 1.1 below, failure rates increase as control systems move toward the end of their lifecycle.

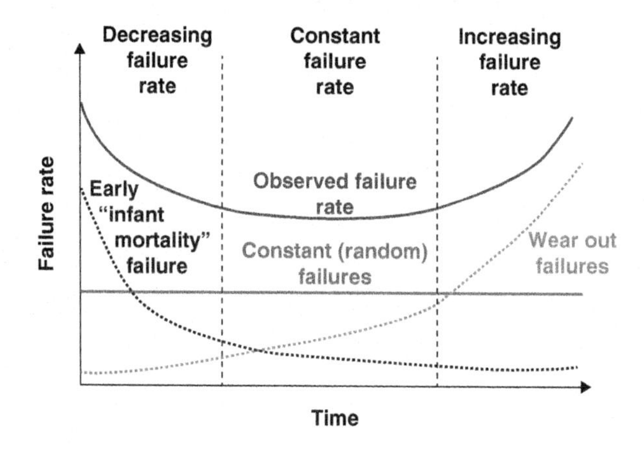

Figure 1.1. Typical control system lifecycle reliability curve.

Control system migration projects take a substantial amount of definition and planning. If reliability has already become a significant issue, the project schedule is likely to be compressed. This often causes deviations from methodical and proven approaches for selecting the right vendor, performing thorough upfront planning, taking full advantage of features and benefits of the new system and developing a comprehensive operational transition strategy.

If control system failures create an urgent project need, project costs also tend to increase. The higher costs are associated with compressed schedules, paying a premium for engineering or vendor resources, and poor project planning. Without proper planning another area that is impacted is the ability to take advantage of improved functionality within the new control system. For instance, alarms in the control system are configured like for like with alarms in the old control system with no thought given to their applicability and effectiveness. This can often reduce the effectiveness of the new control system and translate into a higher ongoing cost of ownership for the system. If your facility is in this situation and there is a push to get the control system migration done quickly to eliminate shutdowns, take time to understand and communicate the potential negative impact on the project.

If reliability issues exist but are not to the point of forcing schedule compression, then you are in a strong position to get approval for the migration project while also following best practices to execute the project. The economics associated with control system failures are more tangible and clear to management when they are experiencing lost production. It moves the argument for the need to migrate a control system from the theoretical to the real. The example below outlines the significant financial impact control system failures can have on a company's operations.

Example

A polymers plant with a continuous process has the control system fail three times in the course of a year. Each failure costs at least one day or 24 hours of production at a normal production rate of 40,000 pounds per hour (PPH). The average price per pound of product is 70 cents. Three days of downtime result in a financial impact due to lost production alone of $2 million.

This does not account for any cost associated with getting the unit back up and running, such as maintenance and operations personnel overtime. This is also unscheduled downtime, which means that the plant is often not able to take advantage of these outages for other maintenance activities that could prevent or delay a scheduled outage at a later time.

The impact of the example above is not isolated to lost production. In today's environment of on-demand scheduling, the entire product wheel can be thrown off when events like the example above occur. If this lost production occurs at a critical point in the schedule when you are making a specialty polymer that is only produced on that specific production line it may result in a missed shipment. This damages your brand. If it causes your customer to miss production targets and shipments to their customers, it may even result in your company losing the customer account.

What is the financial impact of these additional areas? That is difficult to answer and I do not suggest that you try to define any quantitative financial impact for this in your ROI calculation unless it has actually occurred and you have real numbers. However, I would suggest that you build this what-if scenario into your supporting justification documentation to raise awareness that when you start having system failures, there is a domino effect on your business. This message will resonate with management teams who are very customer focused.

PARTS AVAILABILITY OR OBSOLESCENCE ISSUES

As legacy systems get older, vendors reduce and eventually eliminate the availability of certain components whether I/O cards, controllers, consoles, or any variety of subcomponents. Most vendors use a staged approach, but the system cost of ownership immediately begins escalating as parts phase-outs begin. There are a few different approaches that companies typically use to address potential issues with parts availability:

1. Pre-stock additional spares in-house
2. Coordinate with a distributor or third-party vendor to stock additional spares
3. Find a refurbished parts dealer as an alternate supplier.

Unfortunately, parts cost a premium with any of these options. Maintaining excess in-house inventory is inefficient, has tax implications, and is generally not desirable for most companies. Most distributors will charge stocking fees in addition to the elevated parts costs from the vendor. And while there are many reputable refurbished parts dealers the reliability of re-built parts is always a concern.

In addition to the higher system costs, a second issue when parts become obsolete is the potential elimination of expansion capability. Control system capacity limitations can create significant challenges to a company's ability to execute projects needing integration with the control system as highlighted in the example below.

Example

Your plant site is going to build a new unit, which will add 500 points to your existing control system. A last-time buy offering on controllers for your version of the control system was two years ago. Your current controller is near recommended maximum load and will not handle the additional unit. What are your options? There are creative ways to use serial communications from programmable logic controllers and only bring absolutely required points into the controller. There is also an option to eliminate nonessential points in the existing system, but there are likely not many that will be identified. Both of these options are workarounds that just avoid the inevitable conclusion that you need to replace the control system. Maybe the appropriate choice is to use the new unit as a compelling reason to migrate?

Most vendors provide upgrade paths that can help buy additional time operating an older system without requiring complete transitions to a newer system. These may be partial upgrade paths or designed interconnectivity of older and newer control system components. This can be a viable option under certain circumstances. However, when considering these options keep in mind that the vendor is motivated to lock you in with a financial commitment so that when a complete upgrade does occur it will be more difficult to justify migrating to another vendor's system because of the additional investment you have made in the current vendor. If you have no desire to consider other vendors and are planning on moving to your existing vendor's latest platform in the future, then this is an excellent option. One further caution is that once you start mixing components of various generations of products, it generally increases your maintenance costs and makes overall system maintenance more difficult.

If limited parts availability or obsolescence is a part of the issue that motivates the need for a control system migration, there are a few quantitative values that can be included in the ROI. The first recommendation is to capture the differential cost between historical and anticipated escalated parts cost. If a part's price has increased by 25% in a year and the plant purchases roughly five annually, then capture that increase as additional cost associated with keeping the existing control system minus any normal annual escalation pricing.

A second financial indicator to track and capture is the increased cost of adding points to the control system. Cost per I/O is an average dollar amount that represents the cost of hardware, software configuration, and any other I&E and engineering services related to adding an I/O point to the control system. Most controls engineers have a good idea of what this number is for their system. Consider areas where pricing has increased and calculate a new cost per I/O as the system nears end of life. Use your historical cost per I/O for the system as a baseline to do a comparison and capture the differential. This cost per I/O will increase dramatically as parts availability and more importantly, parts obsolescence becomes an issue.

In some cases, the severity of the parts availability issue may be limited to delays in parts delivery. For instance, instead of a standard four-week delivery, a controller that is being phased out may require eight weeks to deliver. While this may seem like little more than an inconvenience, it can have a financial impact in certain situations. If the part is needed for a project longer lead times can usually be planned into the schedule, but for maintenance activities that is often not the case. The only ways around this issue are to pay expediting fees when available as an option or to stock extra components. Both of these approaches again increase the maintenance costs associated with the system. Isolated instances of parts availability issues can be overcome, but when numerous parts within a system become difficult to obtain in a timely manner the practical longevity of the system should be evaluated.

DIFFICULTY INTEGRATING WITH NEWER APPLICATIONS AND SYSTEMS

The early generations of control systems were largely intended to work in isolation as standalone entities. Over the years this philosophy has changed. To varying degrees, modern control systems are interfaced to and integrated with many third-party systems and applications such as those reflected in Figure 1.2.

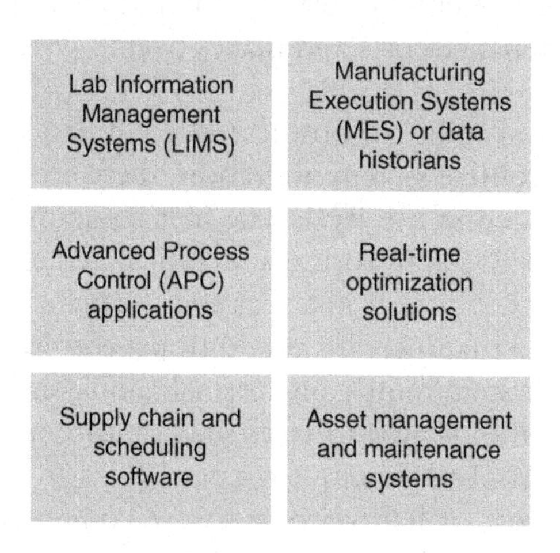

Figure 1.2. Common third-party solutions requiring control system integration.

The architecture and proprietary nature of historical control systems often does not lend itself to seamless communication with these other products. With many older generation control systems the interfaces and/or integration to these other systems is difficult and in a few cases impossible. Brute force methods or customized solutions are often used to enable communications or transfer data. These solutions are both labor intensive and difficult to maintain. Even when data can be successfully passed from the control system, it is often in a less than ideal format. This can reduce the effectiveness of third-party applications and systems decreasing their business benefits.

Defining the ROI associated with interfacing and integration issues is not straightforward. First, document any customization that has been done to facilitate existing communications. If there are maintenance issues or challenges with these customizations, then apply a maintenance cost to them, especially if there is recurring work involved. Second, note any manual activities that take place as a result of system deficiencies, such as the case outlined in the example below.

Example

An inferential model was just developed, which resides on a third-party server. The model predicts several product quality parameters that will be used in an improved control algorithm within the distributed control system (DCS). The expected cost savings resulting from reduced off-quality production losses is $3 million annually. Unfortunately, the model cannot pass predicted value data directly to the DCS because there is no standard communications driver between the inferential model application and the older control system.

Two options are identified as viable solutions. One option is to build a custom driver, which introduces another maintenance point into the system and is expensive. A second option is to have the information sent to the data historian and require that the operator look up the value at some regular frequency. The operator would then have to manually enter that information on a graphic at the DCS console so that it can be used by the control algorithm.

Neither option is ideal, but to quickly get the improved control algorithm in place, you select the option of having the operator manually enter the data at regular intervals. This option requires operator time, distracts from other operator responsibilities, and runs the risk of a manual entry error that can affect plant performance.

In this example scenario above, you would want to capture the costs as well as the risk associated with this manual process in building your justification. If newer control systems offer standard drivers that can securely pass data from the inferential model to the control system, you would also be able to note that as an additional benefit to a migration.

It is equally important to identify any deficiencies in integrated third-party applications and systems that are related to the control system's inflexible architecture and limited communication capabilities. Validate that newer control systems can use standard drivers or other communication methods to exchange information with the third-party system to eliminate the deficiencies. In most cases you will find that they can. If in your situation they do, you have the option of adding these to the benefits of a migration either as supporting information or by including quantitative financial benefits in your ROI calculation. I would caution that unless you can determine a reasonable financial benefit value to claim, which will not be controversial, you may be better off just documenting this as an additional benefit and citing supporting examples.

REDUCED AVAILABILITY OF SUPPORT SERVICES

An often overlooked pitfall of older control systems is the challenge in finding resources to support the system. This is true of both engineering and maintenance support including vendor, third-party system integrator, and in-house resources. The retirement of many of the original contributors to the design, installation, engineering, maintenance, and operation of these older systems results in the loss of significant knowledge, which is not easily recaptured. In the case of vendors and system integrators, they eventually reach a break-over point where it is no longer reasonable to support these systems because there is simply not enough need. Training or maintaining system experts much beyond the end of the product life may not be cost effective. When knowledgeable resources can be found to support aging systems, they are often expensive and demand a premium knowing you have few options. You will want to capture any service price increases related to your system in your ROI as a burden on the total cost of ownership. The example below illustrates this point.

Example

Last year your maintenance contract with the control system vendor was $50,000. The system has gone into the final years of support and the vendor has increased the annual support contract to $80,000 for the same scope of services citing fewer resources available to support the older system as a primary reason. It is not uncommon to see these kinds of increases for services at the end of a control system life.

Let's assume your vendor normally includes a 5% annual escalation in the support contract. In this circumstance beyond that 5% there is an additional increase of $27,500 that is directly related to the late stage lifecycle of the control system. These support costs will continue to increase until the system support services are no longer available. As you build justification for a migration project, you capture the difference between the normal escalation and the current support contract pricing as added cost to continue operating with your existing control system.

To be fair to vendors, their support costs increase as systems near the end of their lifecycle as explained in the example above. However, many vendors also use this as an opportunity to help motivate end user companies to move to their latest system.

When using external resources such as vendor or system integrator engineers and technicians to support near end-of-life control systems, the availability of these resources usually becomes more challenging, meaning you may have to wait longer for their services. When your control system was in the middle of its lifecycle, you may have been able to call and schedule someone to work on your control system within a few days to a week. An older system with fewer resources to provide support may require scheduling a month or more in advance. When the services are needed for projects, you may be able to plan for this, but on maintenance activities you often cannot wait. For example, if you have a workstation fail, you might need a quick response. In this situation, maintenance work will either be delayed or you may have the option of paying extra fees or a premium for an expedited service call.

To account for the increased cost of third-party services, review the pricing for previous projects or maintenance activities. Request a quotation for the same scope asking the service provider to update pricing with current rate. Be sure to explain the purpose so that the service provider does not think it is an active project. Use the differential cost minus any normal escalations to show any increased cost and include this in your ROI calculation. You might also want to ask for an estimate of a similar work scope on a newer control system and include it for comparison purposes.

While issues finding third-party service providers for your control system can be challenging late in the life of a control system, losing knowledgeable in-house resources over time can cause even more difficulties for end user companies. In many older systems, a tremendous amount of coding, graphics scripting, and other customizations were required to achieve the desired flexibility and functionality from the control system. In many companies, these customizations are not well-documented and in-depth knowledge of the control system is isolated to a few individuals. When the resources responsible for programming, maintaining, and operating these systems are no longer available due to retirement, job changes, etc., it reduces the efficiency and increases the cost of most control system related activities. For example, troubleshooting activities can take substantially longer, leading to maintenance inefficiency and longer periods of downtime. When system changes are required, engineering and configuration efforts can also be both time-consuming and costly.

A percent efficiency factor can be used to account for some of these effects in an ROI calculation. Assuming an efficiency factor of 1.0 for the expert resource, the efficiency factor for a less experienced resource may be 0.70, which means a 30% reduction in efficiency, which can be applied to both cost and schedule for control system-related activities. It is an estimated factor, but as long as you make logical assumptions and document them it is a reasonable approach. This efficiency factor can be applied to engineering and maintenance activities using in-house resources and included in the calculation of total cost of ownership.

OPERATIONAL INEFFICIENCY

Understanding the capabilities of newer control systems is essential for identifying operational improvement opportunities that modern technologies can provide. Justifications should not only be based on the costs associated with the weaknesses of the existing control system but also on capturing the operational benefits of newer control systems. My thoughts on migration projects and operational inefficiency are captured very well in the following article excerpt:

> The objective should not be to simply replace and replicate, but rather to innovate. DCS migration is an opportunity to design, implement, and maintain a 21st-century control system that will enable the manufacturer to operate more efficiently and safely; position it for market growth; and firm up its stability for a quarter-century or more. Further, it is an occasion for process improvement and expansion, as well as a much needed opportunity to address the gap between how the plant is currently operated and controlled and how it should be.[1]

The new functions and features available in control systems today along with the emergence of standards and best practices can enable major improvements in operational efficiency compared to older control systems. For example, the automation industry has made tremendous progress in establishing best practices in situational awareness areas over the past twenty years. As our knowledge has grown on how to improve situational awareness, control systems have evolved to incorporate tools to align with those best practices in modern systems.

These new system capabilities combined with the best practices can significantly improve operational efficiency and performance. Better situational awareness helps operators to more quickly recognize and act on abnormal events, which can be safety, quality, or equipment related. Better designs reduce operator distractions so that the operator can focus more time on control of the product quality and throughput. Improved contextualization of information also results in faster and better informed decision-making. The specific impact on ROI includes incident avoidance, production quality and rate improvements, and reduced operational downtime.

Are your company's control room operators as efficient and effective as possible? In most cases, the answer is unequivocally no and it is well-known and generally accepted. Consider defining the costs of operator ineffectiveness where you can identify and claim a realistic portion in your ROI calculation as benefits for a modernized control system.

To capture operator ineffectiveness, first review the operational incidents including safety, quality, and lost production over the past five years. Capture those that were likely due to or contributed to by deficiencies in situational awareness. Quantify these incidents in terms of production losses, downtime, or off-quality product. It is unreasonable to expect that a new control system will avoid every operational issue, but you can claim a reasonable percentage

[1] Matt Sigmon, "DCS Migration: Failure is Not an Option And Doing Nothing is Not a Solution," *Control* XXV, no. 12 (December): 41–42.

from enhanced situational awareness tools such as improved graphics and better alarm handling and management. Be sure to state the basis of your assumption as part of your ROI calculation explanation. The example below illustrates how an operational incident can be used to capture inefficiencies and operator ineffectiveness with an older control system.

Example

An operator is responsible for running the distillation area within a chemical plant. The HMI graphics were designed 25 years ago when the control system was installed. They do not use current recommended best practices for color, layout, background, or alarming. During a thunderstorm, a flood of alarms occurs as temperature deviation alarms are activated. The operator acknowledges the alarms but in the process acknowledged an alarm that one of the column feed pumps has shutdown. Eventually the entire unit is shutdown. The enhanced alarm management capabilities and streamlined HMI graphics available in a modern control system built using current best practices would likely have avoided the alarm flood and made the pump feed alarm more easily identifiable.

Another aspect to consider regarding operational efficiency is reliability and percent uptime. Newer instruments have the capability to send a lot more diagnostic information to the control system helping to quickly identify the root cause of maintenance issues. This enables a preventative approach to maintenance often avoiding or minimizing the operational impact of instrument and equipment problems. Many of the older control systems do not fully support the diagnostic capabilities of newer instruments and equipment. In addition, the internal diagnostics built into modern control systems can help avoid and certainly minimizing troubleshooting time associated with control system issues. In older control systems identifying root cause issues can be difficult resulting in less efficient troubleshooting efforts.

In legacy control systems, the programming languages were often customized and vendor specific, while with newer systems they are standardized, common languages. This enables companies to bring much more consistency to how things are programmed and reduces customization. The result is a simplification of troubleshooting for both maintenance technicians and engineers, which can reduce downtime.

These last two aspects of operational efficiency are not something I would recommend trying to quantify. However, I think they are important points that should be included in the business case justification defining the benefits of control system migration. I would suggest including realistic case examples applicable to your operations as part of your supporting documentation.

SUMMARY

Each organization has unique decision processes when it comes to funding projects. There is no replacement for your understanding and knowledge of your individual organization. When key decision-makers must decide on the merits of a control system migration, they frequently overlook the beneficial aspects and focus on the main points. They consider that the migration may include potential operational downtime, create chaos in the control room during the transition, require re-training of personnel, and cost a significant amount of money.

Common misperceptions and biases regarding control system migrations require controls engineers and project managers do an extremely effective job defining the benefits and ROI of the project to gain support and approval. In this chapter, we have examined some of the common motivations for migrations. As stated earlier, it is seldom a single factor but an additive combination of several of these areas that ultimately builds the successful case for a control system migration. Many of the benefits of a modernization are intangible and difficult to quantify in an ROI calculation. It is important to document these benefits even if not numerically, to build understanding within the organization of your migration project's business value.

Many times, justification efforts are focused on reliability, obsolescence, and other problems with the existing control system. These are valid and are certainly key components of the case for migration. However, as you develop your justification case, also document how improvements in technology will give you an opportunity to improve operational performance. Outline ways to take advantage of new functionality and better tools within modern control systems as they relate to improving your process operations.

Gaining the support of others with a vested interest in seeing a migration occur is a critical starting point. Do not be afraid to incorporate multiple scenarios and justification points into your business case. Ultimately, the decision to approve a control system migration will largely be based on your ability to sell the value to your management team.

Three Key Takeaways

- Include as many key stakeholders as possible early in the process to help build the business case for your control system migration.
- Justification is likely a combination of multiple factors and the ROI is an additive result of these components.
- Do your homework and identify the operational performance improvements that you can take advantage of as a result of new tools and functionality within a modern control system.

2

A Comprehensive FEL

Whether completed as a standalone Front End Loading (FEL) study or included as the first execution step of a full control system migration project, a comprehensive FEL is a necessity. If done well, this planning phase will establish the foundation for a successful project. If done haphazardly or skipped, the execution phase of your control system migration project will be wrought with unforeseen issues that cause schedule delays, cost overruns, and operational challenges. A comprehensive FEL is the most important step of your migration project!

The primary goal of the FEL concept is to push effort forward to the planning and early design phase of a project to identify risks up front when you have a greater ability to reduce or eliminate them. FELs also provide an opportunity to evaluate the feasibility of project strategies before large funding investments are made. There are many variations of FEL processes being used today. Most FEL processes originated from and focus on heavy construction industrial projects, but the concept is a good fundamental engineering and project management practice equally applicable to control system migrations.

FEL processes use a gated methodology with increasing detail and resolution as the project moves forward to the execution gate. A given set of agreed upon criteria must be achieved before proceeding from one gate to the next. These criteria usually involve a combination of the specific deliverables you must have prepared, defined requirements for the detail in those deliverables, and a variability threshold for estimate granularity. For instance, in a three-stage FEL, a Level 1 FEL might require a feasibility study and an estimate with a ±50% variability, while a Level 3 FEL might require a final detailed project scope with a ±10% estimate variability. While some companies use a multi-staged FEL approach for control system migration projects, others choose to consolidate these phases into a single FEL. Either of these can work as long as the process utilized helps define the project details and flushes out risks.

In this chapter, we will examine how to develop a comprehensive FEL specific to a control system migration project. We will begin by evaluating the options for staffing FELs including a review of the pros and cons of each option. Next we will identify and define the key deliverables that commonly comprise migration FELs. Finally, we will discuss some of the important decisions specific to control system migrations that must be made in order for your FEL and subsequent project to progress.

SELECTING YOUR FEL RESOURCES

Determining what resources should staff and execute your control system migration FEL is a critical initial decision that will set the tone for your project. What does the ideal FEL team look like? The answer depends on your organizational philosophy and project details. While you may think that all parties you would consider can do an adequate job of performing the FEL, the differences while subtle can have a significant impact on the quality and thoroughness of your FEL documents.

There are a number of key characteristics as identified in Table 2.1 below to consider and prioritize as you evaluate FEL resource options. I generally use a priority category ranking system as follows: A = Essential, B = Nice to Have, C = Neutral, and D = Not Applicable.

Table 2.1. Prioritization of key knowledge characteristics

Key Characteristic	Priority Category
Understanding of your process or operations	
Knowledge of your specific facility	
In-depth knowledge of your legacy control system(s)	
Familiarity with control system migration best practices	
Knowledge of modern control system features and capabilities	
Extended capabilities for other project requirements (e.g., field construction)	
Sound project management practices	
Experience with communications protocols (e.g., Modbus or OPC UA)	
Background in other areas pertinent to your project (e.g., Wireless or fieldbus)	
Proven and documented FEL work processes	

You may want to add to this list other characteristics that are important to you and apply priorities to these characteristics as well. Identifying the knowledge and capabilities you feel most important for your migration project will simplify the resource evaluation and selection process. It can also help you put together meaningful FEL bid request documents that focus on those characteristics which you have prioritized.

The decision of what resources you use for your FEL is influenced by your process operations, existing control system infrastructure, and overall migration scope. If you have a proven and trusted resource that knows your facility well and is also familiar with control system migrations, then they are an obvious choice. Many companies have a standardized approach to engineering services and obviously you have to work within the parameters of your company's business model. It is also increasingly common for companies to employ a main automation contractor (MAC) commercial model in any number of variations using either vendors or independent automation service providers. Core to the MAC concept is the assumption of responsibility by the MAC provider for the coordination of all aspects of the automation project from the FEL through final commissioning and turnover. This can be an effective strategy for end user companies with resource limitations or a lack of automation system knowledge. However, as a word of caution, there are also negatives to the MAC concept depending on how it is executed. Even with a MAC agreement in place, the end user should be actively involved in monitoring the project and taking part in critical system and service decisions related to the migration.

If you do not already have a resource identified to execute your control system migration FEL, Table 2.2 below lists common resource options along with typical strengths and weaknesses of each. Note that there can be exceptions to some of these general strengths and weaknesses. For example, larger system integrators may have their own construction capability so that the particular scope area is not a weakness for them. Also, keep in mind that a blended combination of these resources can be a viable and effective approach. This blended approach does create some additional challenges with communication and handoffs. However, as long as roles and responsibilities are clearly defined and project management is strong, these challenges can be addressed.

For the most effective evaluation and selection process, using Tables 2.1 and 2.2, you should compare what you deem as critical factors for your project against the strengths and weaknesses of the different resource options to determine the most appropriate resource to staff your FEL. Do not underestimate the value of resources experienced with migration projects. Resources that have extensive experience in these types of projects have insight into what hidden obstacles are common and can proactively help you develop a plan to avoid these pitfalls.

Besides the considerations identified in Table 2.2, there can be other factors relevant to the evaluation process as well. For instance, the location or

Table 2.2. Summary of options for FEL execution

Resources	Strengths	Weaknesses
• Control system vendors	• In-depth knowledge of their control system • Best practices knowledge • Generally sound project management processes • May be able to offer unique hardware or software solutions	• System biases that impact FEL results • High labor costs • Lack of knowledge of other vendor systems • Ability/willingness to work with other vendors
• System integration companies	• Vendor neutral (may still have biases) • Broad knowledge of systems and approaches • Mid-range labor costs • Ability/willingness to work with vendors	• Limited number of experienced resources • May have deficiencies in certain scope areas (e.g., construction) • Depth of knowledge on individual vendor systems
• In-house end user resources	• Knowledge of existing system functionality • Process understanding • Insight into areas of potential issues • Low-capital labor costs	• Workload and lack of dedicated project time • May not be aware of latest technologies, advanced functionality or current best practices • Limited number of resources
• EPC companies	• Generally sound project management processes • Ability to address all facets of the project • Capability to manage labor costs by shifting workload between design and engineering staff	• May lack senior automation-focused staff • Depth of knowledge of individual vendor systems • May have limited software configuration and integration experience

proximity of the resources can impact your interaction with them so that may be one consideration factor especially relevant if your facility is in a remote area. It is difficult to find a resource solution that offers the best of everything, which is why it is important to identify and prioritize the key resource capabilities most vital to your specific project.

IDENTIFYING KEY ENGINEERING DELIVERABLES

Companies can sometimes make the mistake of viewing an FEL as a scope definition when it is actually much more than that. Good FELs should be comprised of numerous deliverables that not only capture the scope of the project, but also define the functional requirements and the standardized approaches that will be required to execute the project. What are the standard deliverables that should be part of your control system migration FEL? The answer varies among companies and is in part determined by how you choose to approach your migration. Table 2.3 below lists what I consider to be a fairly exhaustive set of common deliverables for a control system migration project.

Table 2.3. Common FEL scope deliverables

Category	Deliverable	FEL	Project Execution	Not Required	Comments
General	Scope of work document				
General	Estimate				
General	Schedule				
General	Risk assessment summary				
General	Technical options evaluation				
General	Training plan				
General	Detailed cutover plan				
Controls	Hardware requirements specification				
Controls	Software requirements specification				
Controls	Software configuration specification				
Controls	Control system functional specification				
Controls	Graphics or human-machine interface (HMI) design guide				
Controls	Alarm philosophy/alarm system guide				
Controls	As-is process control network diagram				
Controls	Proposed process control network diagram				
Controls	Control system architecture drawing				

(*Continued*)

Table 2.3. (*Continued*)

Category	Deliverable	FEL	Project Execution	Not Required	Comments
Controls	Marshalling panel design guide (philosophy)				
Controls	As-is marshalling room drawings				
Controls	Proposed marshalling room drawings				
Controls	As-is control room layout drawings				
Controls	Proposed control room layout drawings				
Controls	Loop and logic narratives				
Controls	Cause and effect diagrams				
Controls	Communications plan				
Controls	Factory acceptance test (FAT) plan				
Controls	Site acceptance test (SAT) plan				
Instrument	Instrument specifications				
Instrument	I/O (or instrument) database				
Instrument	Instrument loop sheets				
Instrument	Cable schedule, interconnection and/or wiring diagrams				
Instrument	Cabinet layout drawings				
Instrument	Cable tray routing drawings				
Instrument	Field junction box termination schedules				
Instrument	Field junction box panel layout drawings				
Electrical	Power plan and associated drawings				
Electrical	Grounding plan and associated drawings				

(*Continued*)

Table 2.3. (*Continued*)

Category	Deliverable	FEL	Project Execution	Not Required	Comments
Electrical	Electrical load studies and voltage drop calculations				
Misc.	Equipment installation detail drawings				
Misc.	Plot plan drawing revisions				
Misc.	Piping and instrument drawing (P&ID) revisions				
Misc.	Demolition drawings				

This checklist is intended to help you determine what you want to include as part of an FEL scope, what you want to include in the project execution phase, and what may not be required at all on your particular project. There may also be other unique documents and drawings required beyond this list based on your individual project, but Table 2.3 is a strong starting point for a migration project deliverables list. If you are using a multistage migration approach, you might choose to add columns for each gated phase, as defined by your company process (e.g., FEL1, FEL2, and so on), to the checklist identifying where the production of specific deliverables falls within the migration project execution.

It can be challenging to identify the right time to produce certain deliverables. There are some deliverables that are required at the earliest stages of the FEL to help define the project for estimation and bidding purposes. Other deliverables might require more information that won't be available until later in project execution. The example below presents a common case where a deliverable might be postponed until well into the project.

Example

A detailed cutover plan is a required deliverable for your migration project. You are evaluating whether to produce this during the FEL or later in the actual project execution. At a minimum during the FEL you need to make a determination of whether you plan to approach the cutover hot or cold, whether the migration will be phased over a period of months or years, and address other decisions that can impact the estimate and schedule. You must be able to convey these to those companies bidding on the actual project execution. However, you also know that you are not ready to put together an accurate and detailed cutover plan because there are many design deliverables yet to be developed and your production scheduling six months out when the cutover would take place is not yet available.

One benefit of waiting is that as the wiring details and design is determined the prioritization and ordering of loops to be cutover will become clearer. Also, you expect that the cutover is going to require rolling area outages, and scheduling and logistics will become clearer as the project progresses. You determine that the detailed cutover plan will be delayed until the project execution phase is well under way.

It is not uncommon to combine several of the documents listed in Table 2.3 into a single document or to use variations on these deliverable names. I have provided some further descriptions later in this chapter to better help define the intent of each of the listed deliverables in case you are familiar with a different terminology. There are also deliverables which you might need to have some definition around to develop an estimate or create a bid package, but you may not produce the actual deliverables themselves during the FEL. In these cases, "typicals" are sometimes used to provide a general representation of the deliverable. These "typicals" capture the requirement for the deliverable and put some definition around it. An example outlining a case where "typicals" might be necessary is provided below for your reference.

Example

Logic narratives can be time-consuming and complex to create. For this reason, they may not all be developed during the planning phase of a project. However, to estimate the timing and cost to develop these narratives, the bid package should include some explanation of the complexity and number of logic narratives. Your pipeline operations require six different logic groupings with roughly twenty of each type. To provide an estimate basis and enable an effective bid by engineering companies, you create a single "typical" narrative for each of the six different types and then provide a count of each type. During more detailed engineering as part of project execution, individual narratives will be developed for each.

Another common FEL practice is to include overview versions of some of the deliverables when detailed engineering has not been completed. Overview drawings provide a general idea of the plans but do not include the level of detail that will be required for the final deliverables. This is particularly true of drawings like the control system architecture. If you have not selected a new control system, then you cannot accurately represent the detailed architecture but you can provide an estimate of the number of controllers, number of operator workstations, and general network layout plans.

Defining your FEL process and identifying the specific deliverables required at each stage will provide a clear roadmap as you proceed with your FEL. Just as your FEL process gets more detailed as it progresses, so do many of the deliverables. You might not provide the final deliverable at the first FEL phase but instead use a graduated approach to completing the deliverable as more design details and better project definition become available.

DELIVERABLE DESCRIPTIONS AND CONTENT

In the following paragraphs, I provide general descriptions of many of the deliverables listed in Table 2.3 with the deliverables themselves bolded in the text. Some of the deliverables such as the scope of work, estimate, and schedule are critical project management documents that serve as the roadmap throughout the project execution and warrant a detailed discussion in a dedicated chapter later in this book. The content and nomenclature associated with these deliverables vary across industries and companies, but the explanations below should help you determine both if and how the deliverable fits within your migration project.

We start with deliverables that are part of the general FEL needs for project definition, project management, and the continuation of the project approval process. The **Risk Assessment Summary** is a document that is often overlooked. This document identifies and describes the high-risk areas of the migration. The impact of best and worst case scenarios should be outlined for each risk area and details of the steps being taken to mitigate the risk should be explained. The benefit of creating this document is that it proactively establishes a common understanding among all parties of the potential for issues and the impact of those issues to budget, scope, schedule, and plant operations.

A good risk assessment summary is extremely valuable to management as they evaluate giving final approval to the migration project after the FEL. As shown in the example below a common risk assessment scenario is the decision to reuse the existing marshalling cabinets and insert new marshalling panels inside them on a hot cutover.

Example

You have one particular cabinet for your existing control system with very little I/O currently. I/O in this cabinet will be the first migrated to the new system. In order to do this, some of the I/O in this cabinet will need to be relocated temporarily. This strategy represents a risk of unplanned operational shutdown based on the criticality of a few of these points. The team has determined that the savings in cabinet cost and labor offsets this risk. You use your risk analysis document to present this risk and explain your decision including why it was made and any steps being taken to mitigate the risk.

The **Technical Options Evaluation** is another document that tends to be omitted from many FEL packages. For any key technical decisions within the project, this document captures the options considered and the reasoning behind selecting a particular option. This is valuable during the project execution phase when approaches may need to be re-evaluated. It also provides post-installation documentation so that as personnel change there is a record of the decision logic.

The **Training Plan** is utilized to define the strategy and details for how to familiarize the operations, maintenance, and engineering teams with the functions, features, and daily use of the new control system. It is not developed during the FEL phase, but thought should be given during the FEL phase to expected training requirements and documented in the FEL scope of work. The training plan is a collaborative effort between the project team, the training department, and the individual department whose personnel are being trained. A high comfort level with the new system particular for operations is essential to safe operations, reduced risks of production interruptions, and proper utilization. In most industries, there are regulations related to the training of operations personnel that must be considered as part of the Operator Training Plan.

An essential document that takes significant time to develop is the **Detailed Cutover Plan**. This plan is the roadmap for the details of how the cutover will occur. It is comprehensive and outlines all assumptions, the responsibilities of all parties, and the process of converting points to the new system along with the subsequent checkout and turnover of those points. The most successful plans are detailed to the I/O point granular level and are organized or prioritized by a systematic method. This is especially critical if the cutover is being done hot as issues with a single I/O point can shut down the unit. As mentioned earlier, detailed cutover plans are generally left until project execution, but the general content requirements for the detailed cutover plan are outlined in the FEL scope of work document.

The FEL deliverables related to the controls scope include those requiring design and engineering effort to establish the functional needs of the control system and identify the control system layout, setup, and interfacing. **Requirements Specification Documents** for both hardware and software describe necessary system component constraints or design basis considerations. These are usually brief in nature and may be included as sections in the functional specification rather than as stand-alone documents. The example below illustrates the type of information that might be included in a requirements specification document.

Example

To reduce spares requirements your company has adopted a strategy that all computers be a particular brand. In addition, all process control computers have customized specifications for Redundant Array of Independent Disks (RAID) arrangement and processing speed. These requirements are outlined in your hardware requirements specification documents which are included in your hardware bid package.

The **Software Configuration Specification** is a document which outlines how configuration activities will occur. A control system can be configured in any number of ways to achieve the same result. This does not mean that all of these methods are equally efficient or easy to understand and troubleshoot. This document provides a common process for handling configuration of various control scenarios. For instance, it outlines how and which common function charts should be developed and used. It also covers how any custom coding should be implemented and documented. A good Software Configuration Specification is a reference document that can be used to ensure a consistent approach to configuration of the control system throughout its lifecycle.

A **Control System Functional Specification** defines the required functionality of the control system. It outlines the capabilities and features that the control system must have as well as provides examples of how these are applied. This document will be useful as a checklist in evaluating and selecting your new control system.

The **Graphics or HMI Design Guide** is the roadmap for establishing consistent graphics in your new control system. The determination of how to proceed with graphics is often a major decision point in the project. The HMI design guide provides direction on display related topics such as types of displays, screen layouts, navigation, color usage, equipment and instrument symbols, faceplates, alarms, and general presentation of text and numerical information.

The appropriate time to develop a graphics design guide is the FEL stage because it is not required to be tailored to a specific control system but instead more general in nature. However, some companies approach their HMI design guide differently preferring to make it specific to their particular control system using vendor-specific nomenclature. In these cases, they wait until the control system vendor has been selected before they develop the guide which might be after the FEL. This allows them to use system-specific nomenclature, features, and functionality. Either of these approaches can work, but if you are going to wait until later in the project then you need to determine how to convey graphics requirements to the engineering service bidders during the bid process.

The **Alarm Philosophy Document or Alarm System Guide** describes the purpose of and objectives for the alarm system. It also defines the processes that will be used to meet the objectives. This document covers key areas such as how alarm settings are determined, how alarms are displayed on the HMI, and how they are handled or processed by the operator. There are of course numerous other components that should also be included in a good alarm philosophy as outlined in ANSI/ISA-18.2-2009. Depending on your industry and company requirements, your alarm philosophy document may exist and just need to be provided as a reference. If it does not exist, it may need to be created as part of a separate alarm management project prior to your control system migration.

The **As-Is Process Control Network Diagram** includes a general representation of the I/O, controller and workstation levels of the existing control system, as well as depicting any third-party connections. This diagram should also include network and security access-related devices such as routers and firewalls. The **Proposed Process Control Network Diagram** provides the same information for the new control system. Until a new system is selected, the final network details cannot be determined so the drawing would need to stay at a high level. Even if the drawing is somewhat general at the FEL phase, it is an important part of the bid package for the project as it gives bidders an idea of architecture and component counts for the new control system.

The **Control System Architecture Drawing** is a natural extension of the Process Control Network Diagram and often times they are grouped as a single drawing. The control system architecture drawing contains detailed and labeled I/O cards, controllers, workstations, and third-party connections and interfaces. If you pull the bill of materials from your control system architecture drawing, you should have a complete inventory of your control system equipment including types and quantities.

A **Marshalling Panel Design Guide** includes all relevant details required to design and layout a marshalling cabinet. Some key areas to address include:

- Conduit and cable entry to the cabinet (top entry, bottom entry, or side entry)
- Type, layout and mounting of the terminal blocks
- Wire termination types, wire size, and spare terminations
- Wire and terminal block tagging methodologies
- Grounding, shielding, and lightning protection.

As-Is Control Room and Marshalling Room Layout Drawings should already exist and hopefully only require the confirmation of accuracy and minor updates. These drawings provide physical equipment locations and dimensions within the respective room. Accurately capturing the existing installation is important because transitions from the old system to the new system may require interim equipment installations. These drawings help with the determination of what cutover strategies are feasible with regard to physical space and real estate.

Proposed Control Room and Marshalling Room Layout Drawings indicate the future physical room layout with the new system installed. Until a new system is selected the dimensions of consoles, exact count of physical hardware, etc., cannot be determined. For this reason, these drawings if included in the FEL package are usually general in nature and must be revised with specific details during project execution.

Complex Loop and Logic Narratives can be challenging and time-consuming to create. The narratives normally require input from the end user or a strong understanding of the process as they relate directly to how the unit is controlled. These narratives are typically some combination of block diagrams and text descriptions that capture the functionality of the loop or logic. The narrative must be descriptive enough that a programmer can configure the loop or logic to functionally work as intended. As mentioned in a previous example, during the early FEL stages, "typicals" are often used with an associated count and then narratives are completed for all loops and logic as part of the final engineering effort prior to configuration.

Cause and Effect Diagrams reflect the logical relationships between an event (e.g., a low flow interlock) known as a cause and the resulting system response to the event (e.g., shut down the pump) which is known as an effect. Cause and Effect Diagrams are relevant if your control system migration includes safety interlock functionality, which is normally associated with Safety Instrumented Systems.

The **Communications Plan** identifies all third-party applications, systems, and devices with which the new control system must integrate. This can include complex systems such as PLCs, Safety Systems and Manufacturing Execution Systems (MES) or simple devices such as hardwired shutdown switches. The communications plan designates the communications protocol and defines redundancy levels and any other specialized configuration needs. Typically a basic plan is put together during the FEL as part of the scope definition, but the final detailed communication plan is not completed until the project execution phase.

There are two test plans developed and used during the course of a migration project. The test plans themselves are not required during the FEL phase, but the designation of what test plans will be required is necessary. The first is the **Factory Acceptance Test (FAT) Plan**. This plan includes a point-to-point test of hardware, functional testing of all equipment, and testing of the configuration to some level determined by the specific needs of the project.

The **Site Acceptance Test (SAT) Plan** is also commonly developed during the engineering phase of a project and includes an on-site testing of the system confirming power, grounding, and signal integrity prior to cutover activities. The degree of detail within an SAT depends on the cutover approach and the availability of physical space for staging among other factors. A thorough FEL scope will capture both of these test plans as requirements during project execution.

The instrumentation scope deliverables include a great deal of work that starts with early design effort in the FEL but is not completed until the project is executed. **Instrument Loop Sheets** are used in most facilities as the primary detail drawing, identifying all components of a loop: associated signal, power and ground wiring, terminations, ranges, engineering units, settings, tags, labeling, controller actions, etc. Updates to these are generally not done until the actual project execution, but it is common to include typical examples and estimated counts of individual loop types as part of the bid package for project execution. **Instrument Specifications** may also be required documents if new instruments are part of the project. These specifications provide critical information related to the instrument such as brand, model, calibration information, etc.

The **I/O Database** is a complete list of the I/O in the control system. The fields of the database can vary by company but generally include information such as tag names, descriptors, ranges, engineering units, signal type, alarm set points, I/O addressing, etc. The database should include loop associations as well. For example, two physical I/O points, one analog input and an analog output, comprise a loop. The controller tag associated with these points should be grouped with the individual points so that the software configuration and the physical I/O are linked.

Cable Schedules, Interconnection Diagrams, or Wiring Diagrams are drawings used to outline point-to-point wiring connections between an origin and a destination such as a junction box and a marshalling cabinet. Some companies use one of these diagrams, while others may have all three. Regardless, these drawings are needed for field construction activities as well as development of final loop sheets.

Layout and Termination Drawings for Field Junction Boxes, Marshalling Cabinets, and I/O Cabinets are typically not completed until the vendor selection is made as equipment dimensions and termination details vary among systems. The layout drawings show dimensions and locations of internal components, conduit entry points and sizes, etc. The termination drawings provide a detailed listing of all terminations within the particular box or cabinet. **Field Junction Box Drawings** are only updated if changes to the field instrumentation or field wiring are made. If new field cabling is required, then **Cable Tray Routing Drawings** will also need to be updated indicating how each multipair cable is routed.

The electrical scope of a migration FEL is relatively straightforward but very important to ensuring project success. The **Power and Grounding Plans** including any associated termination drawings are generally updated as part of the engineering effort during the project execution phase rather than in the FEL. However, it is essential during the FEL phase to validate cable lengths, perform load studies, and complete voltage drop calculations ensuring that these are not going to be problem areas later in the project. **Electrical Load Studies and Voltage Drop Calculations** will help you identify the cable sizes and lengths needed as well as determining breaker sizes. If preliminary calculations indicate that you are going to be close to any constraints, a more detailed analysis and plan may be required during the FEL stage.

There are a number of other deliverables that are important to consider as well, which can fall into the civil category or require multiple disciplines to work together. **Equipment Installation Detail Drawings** identify the specific power, grounding, signal wiring, and mounting information associated with an instrument or control device. These will need to be provided for all new equipment including instruments and control system hardware. These drawings are normally provided later in the project during the execution phase as support information for the I&E construction team.

Piping and Instrument Drawings (P&IDs) reflect a process facilities piping, equipment, and instrumentation including how they are interconnected. P&ID revisions may be necessary at some stage of the project if it includes any field instrumentation changes, any nomenclature, or loop tag name changes, etc. These drawings are valuable to third-party engineering and construction personnel in better understanding the process and relative locations of various instrumentation and equipment.

Plot Plan Drawings are scaled drawings of the facility including equipment and building locations, elevations, etc. These drawing may need to be revised to reflect any new construction required on the project such as a new marshalling room or new remote instrument enclosure. They are often included as reference drawings to help orientate outside engineering and construction resources with the layout of the facility.

Demolition Drawings document all instrumentation, equipment, cabling, and other infrastructure that will be removed as part of the project. These drawings are important to include in a complete engineering package along with any related instructions providing demolition details.

Due to the wide variations of how migration projects are executed, the deliverables can be shaped and shifted as needed to best fit your project. Just remember that the intent behind FELs is to shift as much design detail to the early stages of the project as possible to identify and mitigate risk. The deliverables that you produce in the FEL need to help you accomplish this goal.

IMPORTANT FEL DECISIONS

The FEL is the time that critical decisions affecting your project must be made. There are many ways to go about a migration and your strategy will be defined by many factors such as the available resources to support the project as well as the budget, schedule, or timing requirements. In the following paragraphs, I have tried to capture and comment on some of the choices that you will need to make either prior to or during the FEL phase of your control system migration. These are not intended to cover all decisions that you will need to make as part of your project but will hopefully provide you with some initial areas to concentrate on as you define the project throughout the FEL process. The more of these questions you can answer prior to beginning the FEL, the easier and more efficient the FEL process and associated development of FEL deliverables will be.

Do I leave existing instruments in place or use this as an opportunity to complete a re-instrumentation?

A re-instrumentation can be appealing as it is an opportunistic time to upgrade the instruments to take advantage of advanced technologies and better diagnostics to improve overall reliability. There are many factors that contribute to the decision including the reliability and performance of your existing instrumentation. If across the board you have very old instrumentation, then you might be warranted in a wholesale upgrade of facility-wide instrumentation. If you have isolated instruments that are prone to failure or unreliable, then you might look at selectively replacing them as part of the project even though it is really a maintenance issue outside of the project scope. However, the biggest downside to including re-instrumentation as part of your migration project scope is that it can dramatically increase your budget and expand your migration scope complexity.

Do I use fieldbus technologies?

If the decision is made to re-instrument, then a subsequent decision point becomes whether to use traditional 4–20 mA signal wiring or to use fieldbus technologies like PROFIBUS or FOUNDATION™ Fieldbus. These technologies can potentially reduce physical wiring, but more importantly can deliver expanded instrumentation information and diagnostics that can enable more successful preventative maintenance programs and faster troubleshooting. The decision to use these technologies needs to be made understanding that while there are benefits, these technologies are a cultural shift and can be complex to install, maintain, and use to full advantage. If bus technologies are employed, it will be imperative that you fully understand how to deploy these technologies ahead of the project.

What is the most effective wiring methodology?

If you are not going to re-instrument or even if you do change the field device itself but do not use fieldbus technologies, then another key decision is how to approach the wiring. It is easiest to leave the existing cabling in place from the field, but if wiring reliability is an issue this may be the appropriate time to replace cables. If you are going to leave existing wiring in place, where do you intercept the wiring for the new system? The common scenarios are:

1. Instrument wiring routed through field junction boxes, marshalling panels, and then to the control system I/O
2. Instruments wired directly to marshalling panels and then to control system I/O (no field junction boxes)
3. Instruments wired to field junction boxes and then directly to the control system I/O panels (no marshalling panels)
4. Instruments wired directly to control system I/O (no field junction boxes or marshalling panels)

The wiring is most complicated in the scenario where there are neither field junction boxes nor marshalling used. This situation leaves you very few options and can make cutovers without downtime impossible. The best situation is the scenario where there are both field junction boxes and control system marshalling cabinets. This provides you maximum flexibility and also allows you to consider replacement of any individual segment of wiring that might be problematic. In the scenarios where marshalling cabinets are present it is most common to introduce the new system at that point. Field wiring to the marshalling cabinet is typically only replaced if there is a specific known problem.

What marshalling and I/O strategy do I employ?

Real estate for marshalling and I/O can be challenging, particularly in older facilities. If there are real-estate concerns, a decision has to be made about whether additional infrastructure needs to be built, whether there is room for transition staging, or whether a rip and replace methodology will be needed.

Marshalling cabinets and I/O cabinets take up space. In many facilities these cabinets are located in small rooms frequently located physically near the control room. The size of these rooms is generally small and designed with minimal room for expansion. As a result, when you start looking at a migration, you have to make decisions about whether you have space for transition cabinets, whether you plan to use the old cabinets and retrofit, or whether you need to build a new physical space for the cabinets.

It has become increasingly common to see Remote Instrument Enclosure (RIE) buildings used because they allow a completely pre-wired control system from the I/O cabinet to the marshalling cabinet. This approach also allows relocation of the marshalling and I/O rooms to more desirable locations in certain circumstances. The downside of this approach is that it can be an expensive option, requires additional physical site construction, and will require re-wiring of the I/O from the field.

How much do I change graphics?

Graphics are frequently a major source of contention in most control system migration projects. The operators are familiar with the way that graphics are laid out and have adapted to operating from them over time. This generally means that they are resistant to changing graphics. I have been involved in numerous projects as an integrator and vendor where because of the concern about making operators uncomfortable, companies made a decision that we would create graphics in the new system that mimic the graphics in the old system. I have tried to keep personal biases out of this book, but this is one that I am adamant about. Do not constrain graphics in the new system to the deficiencies of graphics in the old system. Too many companies simply want to replace like with like to the extent possible to minimize the re-training of operators.

There have been significant advancements in our understanding of how to optimize graphics for better situational awareness. Take advantage of these best practices and the expanded capabilities of modern control system graphics to improve operations. Will it be comfortable? No, it will require an education process for everyone, but it is worth the investment. During the FEL, give thought to how to optimize graphics for improved situational awareness and document this in your graphics design guide. If a vendor recommends porting your existing graphics to the new system, unless you have just recently completed your graphics I recommend avoiding this approach. First, it seldom works as advertised and can cause additional work. Secondly, it prevents you from fully embracing the technical enhancements of the new control system.

Do I need to assess or reassess my alarm handling and alarm management?

Another aspect of improved situational awareness is better alarm handling and alarm management. Alarm management has become an increasing focal point for many industries and organizations. A control system migration project provides an opportunity to revisit alarms and alarm settings. This may or may not be something you want to roll into the control system migration project.

If you have not done an alarm rationalization, do not have a current alarm philosophy document, or have not been managing your alarms on an ongoing basis, you should definitely plan to address alarms prior to configuring your new control system. Unfortunately, once nuisance alarms are configured in a control system, it can take a long time to get them identified and removed if you do not have an effective alarm management program.

I recommend handling alarm management as a standalone project. If it is incorporated into the control system migration, it needs to be as early in the project engineering effort as possible. If a rationalization effort is required, it will involve a number of resources from operations, maintenance, and engineering so determining if this is needed is critical.

What is the best cutover approach?

The decision of whether to perform a "hot" cutover where the transition is on a live operating unit or "cold" where the operating unit is down while the cutover occurs is driven by many factors. One consideration is the amount of scheduled downtime that a facility generally incurs. If there are windows of opportunity built into the operating schedule, then you should plan to take advantage of these.

A second consideration is the risk associated with a "hot" cutover. What is the likelihood that issues with a single point would bring down the unit? What are the implications of an unscheduled outage that occurs as a result of a hot cutover issue? It is also important to understand the resource implications of each scenario. If one scenario requires additional I&E or operations staffing and overtime while the other does not, that may be a factor in the decision as well. The final decision on the cutover method should not be made until the detailed cutover analysis is done as part of the project execution, but a preliminary plan can be evaluated and documented during the FEL phase of the project.

SUMMARY

In summary, FEL efforts are essential to project success. FEL processes vary greatly among companies. Some use a multiphase FEL approach where the project is evaluated at the end of each defined phase to determine whether it passes to the next phase. Others choose to use a single FEL and then incorporate the remaining engineering effort into the project execution phase. Regardless, define what deliverables are needed for your FEL to get you to the level of scope granularity and estimate accuracy that makes you comfortable moving forward with the project.

Selecting a team of the right resources to develop a defined and complete set of deliverables will help provide a successful control system migration FEL. Do not be afraid to explore technical options and different approaches as part of the FEL exercise. It is through these investigations that you can often uncover the hidden obstacles to your migration project's success. Identifying and addressing risks in the FEL stage will reduce the potential for these areas to negatively impact the project execution phase.

Three Key Takeaways

- Understanding your specific project requirements and prioritizing key needs will assist you in selecting the best resource option to staff your control system migration FEL.
- Answering project strategy questions and making key migration-related decisions either before beginning or during your FEL will facilitate a more efficient FEL and subsequent project execution.
- Your project success is dependent upon a well-executed, comprehensive FEL that clearly defines the details of your control system migration, identifies high risk areas, and implements effective mitigation steps.

3

Bid Specifications and Vendor Selection

Bid specification documents and vendor selection are parts of the control system migration project process that are often not given sufficient attention. The development of documents that will be used as the basis of vendor bids for the control system and the engineering, procurement, and construction (EPC) services must be thorough and of the highest quality if you expect accurate, quality bids in return. It is essential to clearly define the content you expect bids to include as well as the way you prefer information to be formatted and grouped so that you can easily evaluate and compare bids against your key selection criteria.

I suggest that you bid the control system hardware and software separate from the EPC services. Control system vendors generally do not like this approach because it prevents them from shifting costs between various areas of the project, but it gives you the most accurate reflection of control system costs. It also enables you to bid the services to different sources such as system integrators or EPC firms and get an accurate comparison value for the same scope. Many vendors argue that they can provide better pricing if the project is bid as a whole. If a vendor wants to propose discounting that is related to a bundling of services, let them do so as an option after they have complied with your request for separate system and services bids.

The two primary goals of the bid request process are as follows:

1. To obtain accurate pricing information and project schedules based on the bid scope.
2. To get the same scope bid from each bidder so that an accurate comparison and bid evaluation can be done.

To accomplish these two goals, documentation issued to bidders must be complete and clear leaving little room for individual interpretation. This includes not only the documents that are technical in nature, but also the bid instructions themselves. For example, if you want the bids organized in a certain way, then that must be clearly defined in the bid instructions. Lack of quality and thoroughness in bid request documents lead to lack of quality and thoroughness in bids resulting in change orders and escalating project costs as your migration progresses.

In this chapter, we examine what comprises a control system bid package for both the control system and the EPC services scopes. We review how to establish and prioritize key vendor selection criteria and evaluate bids with an objective process that shifts emotional choices to logical decisions. We also identify some of the common challenges with the bid process and review ways to avoid these issues.

CONTROL SYSTEM

The relationship between control system vendors and end users is complex. At times it is a partnership and at other times an adversarial relationship. Both parties recognize that in the long run, they need one another to be successful. The selection and purchase of a control system are a long-term commitment that cannot be easily undone. For this reason, it is important that you understand what functionality you require in your new control system and have a defined process for evaluating how a control system complies with your needs.

One of the biggest challenges that many end users face is how to take the emotion out of the control system purchasing decision. I have dealt with a number of companies that struggled to do this. The anecdote below is a case where the loyalties of individuals based on their experiences with the control system vendor risked negatively impacting their migration path decision.

Anecdote

I was asked to help a refinery scope and bid a control system migration project. The existing control system had been installed for over 20 years and there was a long history with the vendor. The controls engineer was frustrated with the vendor who he felt had been price gouging them for years on maintenance support and spare parts pricing. He also felt that because of some recent personnel changes with the sales team that the vendor was much less responsive than in the past. The controls engineer had formed relationships with a new vendor and was convinced that their control system was the best on the market at the time.

The operations manager on the other hand had grown up with the existing vendor. It was all he knew and he was convinced that the upgrade to the existing vendor's newer system was the right choice. His primary goal was to minimize the impact of any control system changes to his operations team. Both the controls engineer and the operations manager had already made their decision on the control system of the future before either had done any evaluation. This meant that I had the responsibility of walking them through an objective selection process.

The first step was to get everyone in the same room and then ask them to list all of the criteria that they felt were important in a new control system. We then reviewed these criteria and reached a common agreement on how to prioritize the list. This process can initiate long discussions and take considerable time, but they eventually reached a consensus. This prioritized list of criteria became the score sheet that was used to rank each vendor and eventually reach a decision on the control system of the future.

While there remained some subjectivity in the process, the evaluation of vendors against each of the key decision criteria became much more objective and numerical. In the end, they were both happy with the process. The operations manager actually pulled me aside at the end of the vendor evaluation meeting and complemented me on the objectivity of the process having recognized that we reached a logical conclusion rather than an emotional decision.

The first step in selecting a vendor is accurately defining what you need in a control system. This is done through a combination of functional specifications, requirements specifications, bid instructions, and a decision criteria matrix. In the sections that immediately follow, we explore the purpose and content of these documents in more detail.

The Control System Functional Specification

As mentioned in the previous chapter, the control system functional specification defines the required functionality of the overall system. It also outlines the capabilities and features that the control system must have and provides examples of how these are applied. I start with this document because it establishes the goals for your control system. The requirements specifications then provide additional details on system and software factors that support achieving these goals. It is not uncommon to see the hardware and software requirements included as sections in the functional specification. The decision of whether to have separate documents or roll them into one is driven by company philosophy, system complexity, engineering costs, etc. For the purposes of clarity, I have kept them separate in this book.

Table 3.1 below reflects a general table of contents that can be used as a starting point for a control system functional requirement specification and includes a brief description of the content of each section.

Table 3.1. General control system functional specification table of contents

Control System Functional Requirement	Summary of Content
Platform	Defines the operating system and general server requirements for processing speed and performance, architecture flexibility, etc.
Controllers	Covers expectations of the controller with regard to processing speed and capacity, continuous or batch handling, addressing, standard control algorithms, and response during power interruption.
I/O	Defines the types of I/O cards desired along with point densities, signal types, power requirements, area classification ratings, etc. Also covers wireless or remote I/O requirements.
Redundancy and fault tolerance	Provides requirements for redundancy at all levels of the control system including I/O, marshalling, cabling, power supplies, controllers, and operator workstations. Additionally, presents acceptable fault tolerance strategy and switchover timing requirements.
Power and grounding	Outlines what components of the system are sourced with what voltages. Also covers expectations regarding fusing, wired versus plug-in, etc.
Consoles/ operator stations	Designates acceptable response times, password levels, interaction compatibility (mouse or touch screen), and other standard operational and maintenance functions.
Third-party interfaces and communications	Documents the acceptable and preferred methods for communication to third-party systems.
Security	Defines the desired levels of security, the password protection requirements, audit trail expectations, and acceptable methods for remote connectivity.
Engineering configuration	Covers what is expected for the standard engineering capabilities of the system including the preferred configuration approach for hardware, points, logic, and graphics.
Alarm handling	Outlines alarm types, levels, priorities, logging, and recall.
Historization and trending	Designates the requirements surrounding how point information is collected, stored, and recalled. This includes the frequency, method (time or event), configurable data compression settings, etc.
Reporting	Defines the expected standard reports, ad-hoc reporting configuration requirements, initiating trigger options (upon request, event initiated, or application initiated), and all formatting specifications.
Diagnostics	Covers the minimum diagnostics that should be standard within the control system such as power failures, I/O card failures, I/O point failures, I/O cabinet fan failures, etc.

Some of the requirement areas in Table 3.1 may be more or less important to your specific control system, but should all be addressed in some way to help vendors determine how to propose the most effective system design for your needs.

Hardware and Software Requirements Specifications

When you have completed a functional specification that designates what you want the system to do, the individual requirements specifications then provide additional detail on how to do it. More specifically, they detail any design requirements or constraints with which the control system vendor may need to be aware. The information in a hardware specification includes any company standards for brands, models, memory capacity, and processing speed related to computer hardware for operator consoles and engineering workstations. The environmental conditions that the equipment will be subject to are outlined in the hardware specification as well as any area classification requirements. The specification also covers other areas such as required physical sizes for operator and engineering workstation monitors.

The software specification documents the capabilities that a control systems engineering and operations software must have and outlines the approach that will be used for the configuration of the control system. For example, you might include a list of any standard function blocks and control algorithms which you expect to be available off-the-shelf in the control system software. The software specification also details how to program any customization including how the programming should be structured and documented. The benefit of a good software specification is that it facilitates consistent programming practices when multiple programmers will configure your system. However, guard against being too rigid and detailed in the specification or programmers will lose the flexibility to maximize the efficiency of your configuration.

Control System Bid Instructions

The control system bid instructions communicate vital information about your expectations regarding the content and format of bids from the control system vendors. Having been a part of control system migration proposals and bids as an end user, system integrator, and vendor, my conclusion is that the bid process presents challenges and frustrations for all parties. One way to reduce these frustrations is to establish clear expectations up front.

The bid instructions document is the foundation of your bid package. This document references all other documents included in the package that the control system vendor must comply with such as the functional specification. You should clearly state how the vendor should identify any areas with which they are not in compliance. The bid instructions also detail the format which you want the vendors bid including breakdown by hardware and software with any appropriate subcategories. Including a bid estimate form with the desired breakdown is a common practice to help facilitate this.

Good bid instructions summarize the overall I/O count and include an estimated breakdown by signal type. By including this detail here, you are promoting a common bid basis. If you leave it up to vendors to go through the I/O index to determine point counts, you are likely to get some variations on I/O counts. This will make it more difficult to compare bids. You should also include a statement as to the percentage of spare I/O capacity by type and possibly by controller that you will require so that this can be included in the bids. In greenfield projects, this is easy because you can typically provide an across the board spares percentage. In brownfield operations, it can be more challenging as you might have variable spare capacity needs by process unit or area.

Bid instructions also outline the number of workstations you require and any furniture that you expect control system vendors to either provide or fit equipment into. Your expectations regarding software licensing should also be explicitly stated in the bid instructions. Keep in mind that vendors have different license philosophies, which can impact the bid process. For instance, some control system vendors have the ability to reduce their bid by consolidating or minimizing licenses. While technically they may comply with your bid request, you may be left without the full capabilities or flexibility that you expect and will soon be purchasing additional licenses.

Shipping and freight costs can be an additional project expense often overlooked. I recommend that you ask control system vendors to either include these costs in their bids or provide you with an estimate of these costs. Dependent upon the point of origin, these can be significant costs. Along these same lines you will want to clearly define your system warranty requirements and specifically identify when the warranty begins (e.g., upon startup). Estimated delivery schedules should also be a part of control system vendor bids so that you can effectively plan these into the overall project schedule.

There are also several other areas that should be addressed in the bid instructions. Any acceptance testing or startup activities that you wish to have the vendor participate in need to be clearly stated. You will also want to state what training support you expect the vendor to provide as well as any required documentation such as technical manuals.

As part of the control system bid package, several attachments listed below are commonly included:

- Functional specification
- Hardware requirements specification
- Software requirements specification
- I/O index
- Marshalling panel design guide
- Overview system architecture drawing
- Control room layout drawing
- Bid estimate form

The quality of your bid instructions document will directly impact the quality of the bids you receive so spend adequate time preparing this document. A good bid instruction document supports an evaluation process that is clearly defined, fairly evaluated, and equalizes the playing field.

Decision Criteria Matrix

Earlier in this chapter, I cited the example of the emotions in the control system vendor selection process and how the use of a decision criteria matrix and scoring system can bring objectivity to the selection. In this section, we walk through how to build a decision criteria matrix, how to rank and score vendors, and how to make your final vendor selection. There are numerous selection methods such as the Pugh decision matrix, Kepner–Tregoe decision-making method, and the quantitative selection matrix. There are slight variations in the details of these methods including if and how criteria weighting and prioritization are accomplished. The process I recommend below is somewhat of a hybrid of these methods closely aligned with the Pugh approach although using a different approach to weighting and scoring. I recommend that you evaluate all of these approaches and use the one that makes the most sense to you.

The first step in developing the decision criteria matrix is to gather a group of stakeholders for a brainstorming session to create the matrix. The team should include at least one representative from the each of the following groups: controls engineering, operations, maintenance, and purchasing. In the case of operations and maintenance, I would recommend several people join the team including at least one manager and one technician from each of the disciplines. There are other groups such as IT that may need to be included depending upon how your company divides responsibilities and whether they have input into the software usage and system architecture on the control system side.

Once your team is assembled, a meeting should be scheduled to brainstorm the key characteristics that the team desires in a control system. What granularity should you use for this process? My experience has been that the major areas fall out rather quickly and the individual criteria can easily be categorized under those major areas. You will see this reflected in the example criteria matrix in Table 3.2.

To streamline meeting time you might bring a criteria matrix that is already prepared and have the team add to or remove from the list as appropriate. My only caution with this approach is that it can narrow your focus and prevent some of the creativity that comes from a free-form brainstorming session. Table 3.2 below can be used as a starting point for your decision criteria matrix.

Table 3.2. General control system selection decision criteria matrix

Criteria	Priority	Vendor 1	Vendor 2	Vendor 3
HMI Graphics				
Configurable layout				
Navigation				
Usable standard faceplates				
Alarm handling & alarm management				
Real-time trending				
Reporting				
Personalization by login				
System Design				
Architecture flexibility				
Scalability				
Operator workstation (console) design				
Controller design				
I/O module layout and design				
I/O density				
Supports fieldbus technologies (profibus or foundation fieldbus)				
Supports standard communication protocols/interfaces to third-party systems, applications and devices				
Maintenance and Reliability				
Standard diagnostics				
Additional configurable diagnostics				

(*Continued*)

Table 3.2. (*Continued*)

Criteria	Priority	Vendor 1	Vendor 2	Vendor 3
Ease of software and hardware upgrades				
Redundancy design				
Comprehensive security features				
Vendor technical support capability				
Quality and accessibility of technical documentation				
Engineering				
Ease of hardware configuration				
Software platform				
Ease of point configuration				
Ease of graphic configuration				
Minimal required customization or scripting				
Allows flexible tag naming				
Flexibility to customize in needed areas				
Minimal system configuration required for standard interfaces				
Commercial				
Total cost of ownership (based on 10-year evaluation)				
Initial system pricing				
Agree to 5-year maintenance support contract				
Standard delivery				
Industry references				
Worldwide installed base				
Local country installed base				
Company installed base				
Long-term commitment to product (roadmap plan)				
Vendor cooperation/bid responsiveness				

Key. Priority factor = 3 means the item is the highest priority and required; priority factor = 2 means the item is highly desired; priority factor = 1 means that the item is nice to have but not essential; priority factor = 0 means that the item is irrelevant and not considered valuable to the assessment.

Some of the criteria above may not be relevant to your specific company's needs. Take for example the similar criteria of industry references, worldwide installed base, local country installed base, and company installed base. You likely aren't concerned about all of these and will choose one or two of these as evaluation criteria. Maybe industry references and company installed base are most important. On the other hand, may be local country installed base is the only one that is relevant to your particular circumstance.

All of these criteria are flexible and you should make sure that the criteria used for your evaluation are meaningful to your particular project. Once the specific criteria are listed, the challenge becomes prioritizing them in a holistic manner. It is natural that the engineer is going to want to put the highest priority on the engineering criteria, and the operations manager on the operations criteria. If everyone takes this approach, then every line item will be the highest priority and the matrix will have little value. The group has the responsibility for talking through the impact of each criterion to overall company cost, employee efficiency, equipment reliability, and safety. This process should help the group reach a consensus on the priority of individual criteria. It is not necessarily an easy exercise, but it is an extremely valuable exercise because at the end of it, the company has a common understanding of their overall control system priorities.

One of the most common mistakes that I have seen companies make in the bid evaluation process is using initial cost as opposed to total cost of ownership as a decision criterion. All control system vendors have ways of shifting their costs in efforts to make their initial bid the most appealing. For instance, some vendors are well known for reducing initial system cost and expecting to make this up through annual support agreements and future system pricing adjustments.

Many end user companies consider this a questionable business practice, but unfortunately it is partially an outcome of many end users not having a good evaluation and decision process. If end users select a system based solely on initial cost, which unfortunately has been done more than once by a zealous purchasing agent eager to lower cost, then control system vendors are forced to adjust their pricing to try to win the deal. Make no mistake that every vendor must make up that money somewhere or they won't be in business long. Some control system vendors with broad instrument, electrical, and control system portfolios are willing to break even or take a loss on migration projects if they view the pull-through value of the account as significant enough. The scenario outlined below presents one example of this.

Example

One control system vendor being considered for your project also has a significant offering of instrumentation and valves. They are aware that a new production line is planned for your facility in the next five years. The vendor has had limited success in getting their instrumentation into your company in the past. If they win the control system bid at break even, they believe it will help their chances of getting the instrumentation and controls on the new production line. So, over the short-term, they bid low on the control system migration project confident that they will be better positioned to win instrumentation opportunities on the future expansion with their control system in place. They consider this a worthwhile investment and expect to gain back the revenue on future pull-through business.

Many end user companies struggle to determine where a vendor should be in the product lifecycle and how to reflect and rank a system's lifecycle status in the decision criteria matrix. If vendors have a control system that has recently been introduced or significantly updated the user might be hesitant to be early adopters. Most end user companies want the new product or new release issues worked out before they adopt the product. For this reason, end users often lean toward a very proven control system. However, at some point a system with longevity becomes self-defeating to your modernization efforts. The anecdote below illustrates this point.

Anecdote

In one particular situation, I was responsible for recommending a control system upgrade plan to the plant management team as part of a control room consolidation project. At the time, the plant's existing vendor had just gone through merger and acquisition activity. The future plans for the plant's particular control system were uncertain. After evaluating several options and scenarios, I recommended simply upgrading consoles to a newly introduced version of the operator workstations that was in Beta testing. This option minimized investment while addressing the reliability issues with the existing system.

The management team was concerned because of the lack of confidence in a long-term commitment by the vendor to their particular brand of control system. They were not sure they wanted to invest any money in a system with an uncertain future. We had the vendor's product management group visit with plant management. The vendor re-assured them of their commitment to the brand and talked about their support plans regardless of the product direction. It reduced the concerns of the operations manager and the plant manager so we moved ahead with the project as planned.

Unfortunately, it was only a short time later that the vendor began selling another control system as their primary product offering. While the vendor continued supporting the plant's systems for another seven years, it was no longer the focus of their organization, development initiatives were slowly phased out, and the end user was left with a control system that had a limited future shortly after they purchased it. The plant eventually upgraded the entire system to the same vendor's new control platform.

In the example above, I believe that the vendor team sent to discuss the longevity of the control system acted in good faith based on the knowledge they had and could share at the time. The point is simply that finding the sweet spot of a control system lifecycle is not easy. You should do your due diligence but in the end recognize that the best control system choice you can make today may not be the best choice tomorrow. As long as you have improved your control system then you have made progress.

When end users evaluate vendors, it is natural to use the relationship with the sales person as a metric. I always warn against putting too much weight on this relationship. The reason is that if you have a really great relationship with a specific salesperson, there is no guarantee that this same individual will be your salesperson six months from now. People frequently leave companies, get promoted, or change account territories. For this reason, it is important to recognize the fluidity of these relationships. Don't discount a good salesperson that is responsive to your needs, but also don't let that overinfluence your decision when selecting a control system.

Selecting a Control System Vendor

After the control system bids are received, the same team that developed your decision criteria should meet again and review all bids. The team should collectively reach a consensus on how each vendor's bid complies with each criterion and record the individual score of each vendor for the respective line item. Again an alternate approach is to take an initial pass at scoring the control systems prior to the meeting to save time. The group would then review, discuss, and comment during the meeting making any necessary adjustments.

To score a vendor on each line item use a 0–3 scale as follows:

3 = Exceeds Expectations, this is a vendor strength and they have the capability to do meet all requirements and then some.

2 = Meets Expectations, the vendor has the capability to meet the minimum requirements.

1 = Falls Short of Expectations, this is a vendor weakness and they cannot meet the minimum requirements.

0 = Does Not Comply At All, the vendor has limited or no capability in this area.

Once the team has gone through this process, each line item score for the vendor is multiplied by the priority factor for that line item determined previously and then all line item scores are added to give each vendor's final total. The highest total score represents the vendor who best meets the overall needs as defined by the company. However, keep in mind that if that vendor falls short in a critical high priority area, they may still not be the best fit. The team needs to review all of the scoring and confirm that they can accept any areas where the vendor may have deficiencies.

In Table 3.3 below is a scoring example with final tallies for the maintenance and reliability section of a vendor selection evaluation.

Table 3.3. Example vendor analysis

Criteria	Priority Factor	Vendor 1	Vendor 2	Vendor 3	Total		
					V1	V2	V3
Maintenance and Reliability							
Standard diagnostics	3	1	3	2	3	9	6
Additional configurable diagnostics	1	1	2	1	1	2	1
Ease of software and hardware upgrades	2	2	2	2	4	4	4
Redundancy design	3	0	3	2	0	9	6
Comprehensive security features	3	2	2	3	6	6	9
Vendor technical support capability	2	2	3	2	4	6	4
Quality and accessibility of technical documentation	1	3	3	3	3	3	3
TOTAL SCORE					21	39	33

Based on the example analysis of the maintenance and reliability section Vendor 2 would be the best selection. While this methodology is not perfect and does include some subjectivity, it promotes a logical rather than emotional-based discussion, provides a numerical score, and formally documents the evaluation process for future reference.

ENGINEERING, PROCUREMENT, AND CONSTRUCTION SERVICES

The Engineering, Procurement, and Construction (EPC) services are every bit as important to the success of the migration project as the control system itself. Engineering, Procurement, and Construction project responsibilities vary greatly among companies. For instance, many companies handle all of the procurement activities themselves so the scope is really engineering and construction. These variations do not impact the bid and evaluation process that is described here. The term EPC service providers in this section refers to the function rather than a specific type of provider and may be an EPC firm, system integrator, control system vendor, or in-house resources.

The bid process, identification, and prioritization of decision criteria, and vendor ranking and selection for the EPC services scope is largely the same as it is for the control system vendor. I will not re-hash the details of the process in this section. Instead we will focus on those unique aspects of the EPC services portion of a control system migration project that can create a more complete bid scope and review some of the common practices that can skew a bid evaluation.

Requirements Definition

It is important to make sure that the services bid requirements are complete and not just isolated to the technical details of the project. Numerous bid requirements related to the EPC services are often overlooked in bid package documents leading to change orders during the project execution phase. Table 3.4 below is a list of areas commonly omitted in bid requirements documentation leading to misaligned expectations and frustrations between end users and bidders.

Table 3.4. Key areas to define in bid documents

Project management responsibilities	Meetings	Project reporting
Cutover management	Change order/RFI management	Company processes and procedures
Drawing and document checkout	Deliverable review cycle	Demolition requirements
Final project documentation	Travel and living expenses	

Project management responsibilities are often not explicitly stated in scope documents. End user companies may have expectations that EPC services bidders will include project management as a necessary component of the project. However, if no guidance has been provided, then it will likely be only minimally included if at all. To avoid conflicts related to project management responsibilities on your migration project include a detailed list of what you expect the EPC services vendor to provide such as organization, planning, and attendance at status update meetings.

The frequency, duration, location, and attendance requirements for meetings are all areas you should specify. Meetings may be held weekly or the frequency may change during different aspects of the project. For instance, during the engineering effort, bi-weekly meetings may be acceptable, but during cutover weekly or even daily meetings may be required. The duration of the meetings should be specified as well as where they are held. Define who is required to attend and whether onsite attendance is required or remote attendance by phone or through a web-based solution is adequate. All of these factors impact the bidder's costs and need to be captured as part of the required EPC services scope during the initial bid process.

Project reporting is another area where expectations can vary greatly between bidders and end users if not detailed. You obviously want to define the expected frequency of reporting, but in addition you want to state the requirements for report content. The simplest reports may only provide a general status update, while more complex reports may include earned versus spent tracking, etc. The reporting requirements are in part influenced by the type of contract it will be. For example, if it is a fixed price bid, then earned versus spent reporting may not be appropriate. Also as part of reporting, do you expect schedule updates? If so you need to include how often schedule updates are expected and what software tools should be used to manage the schedule.

Details related to cutover management are often too loosely defined in bid documents as well. During many cutovers, work days are extended beyond a normal work day and you will want to specify what type of onsite coverage you require so that bidders can account for any required overtime in their bids. Also, what type of regular coordination activities with operations and maintenance should take place during the cutover? If you expect daily cutover meetings, then EPC service bidders would need to include additional time and budget to prepare for and attend these meetings.

Establishing common expectations about a change order management process in your bid documentation is also wise. How do you want to handle activities that are considered scope changes? What documentation and approval process will you require? A common method is notification via a written Request for Information (RFI). The content of an RFI often includes descriptions of the proposed work, why it is considered outside of the scope, why it is needed or required and the expected impact to schedule and budget. Some companies view RFI approvals as permission for the service provider to proceed, while others won't allow work on the specific item to begin until the change order paperwork is finalized. To avoid confusion, be clear about your entire RFI or change order management process in your bid documents.

Also clarify in your bid documentation any company processes or procedures with which the bidder will need to comply. Many operational and maintenance policies impact the time it takes to complete field engineering and field construction activities so it is important to ensure that bidders are clear on these procedures upfront. For instance, if you have a work permitting process that requires a bidder to wait for one hour after shift change for sign-offs each morning then EPC services bidders will need to account for this time in their bids and schedules.

The process of checking out drawings and documents for revision should also be addressed in bid-related documents. This drawing request process can sometimes have a delayed turnaround and the bidders need to factor this into their estimate and schedule. This is particularly true of any older drawings that have not been converted to electronic format. The form to be used to make the request should also be explained. For example, some companies are okay with email request to the document control or drafting department head, while others prefer a more formal notification system that can be more easily tracked.

When the EPC provider submits a document, drawing, or other deliverable to you for review during the project, how long can they expect to wait until you return it? Expected review cycle times should be clearly stated in the bid package as they directly impact the project schedule. Review cycles for EPC service providers can be a source of great frustration. Keep in mind that long review cycles extend schedules and do increase project costs. If you don't include this in the bid package, most service providers will assume a review cycle of one week and then submit change orders if the process is taking longer and negatively impacting the schedule.

Demolition activities are another area inadequately addressed in many bid specification documents. While we included demolition drawings in our control system migration deliverables table in the previous chapter, there is often also a need to provide additional demolition instructions in the scope of work or in a separate document. In your bid documents, clearly establish who is responsible for equipment removal and where the final location of removed equipment should be. If it is to be taken offsite, environmental considerations will also need to be covered. All of these are factors that can impact the EPC services budget and schedule.

Clearly state in your EPC services bid scope what you expect bidders to include in the final documentation package at turnover and what media are acceptable for this documentation. In the past, multiple sets of large project books with hardcopy drawings were part of standard delivery. Today, most end user companies are fine with electronic copies of all materials. Be sure to specify your requirements for file formats and versions, number of copies, etc.

Travel and living (T&L) expenses can also be a source of disagreement during the project depending on the type of contract you sign. If T&L expenses are only estimated for the project or are to be paid in addition to the base bid then clearly establish the parameters of what T&L charges are acceptable with your bid documentation. This is particularly important if you select an EPC service provider that is not local to your area or if you are in a remote area and travel will be required by any outside resources working on your project. In addition to stating the types of T&L charges that are acceptable, you will want to indicate the markup percentage you are willing to pay. Service providers do have an administrative fee associated with T&L expenses so a markup percentage of 10–15% is standard.

Clearly defining all aspects of the EPC services scope during the bid process will benefit both the end user company and the EPC services bidders. Detailing these nebulous areas outlined above in your bid documentation will help establish common expectations and result in bidders providing more accurate pricing and schedules. It will also minimize the change orders related to these areas during project execution.

A Complete EPC Bid Request Package

The content of an EPC services bid package for a control system migration should provide a clear scope of work for bidders and give them direction on how you want their bids organized. Any areas where the details are not fully identified need to be explained and quantified enough so that a firm bid basis is established. EPC service providers whether EPC companies, system integrators, vendors, or in-house resources have the goal of appealing to you as the most attractive option bidder and winning the project. Keeping this in mind, recognize that they will bid exactly what you have told them to through your bid documents as explained in the anecdote below.

Anecdote

I have written numerous control system migration proposals during my career. Earlier, I would throw the kitchen sink into the proposal. If a client had not specifically outlined project management meetings and reports, I would still include regular meetings and reporting as a part of project management best practices. I was quite disheartened after not being awarded a couple of projects that I had expected us to win.

My boss at the time sat me down and taught me the invaluable mantra "Bid the Paper." In the simplest terms, that means my proposal includes exactly what you requested in the bid documentation and nothing more. If you unintentionally leave something out of your bid request, or assume that I will include it but don't specifically outline it, don't expect it to be a part of my bid. Also if you don't define activity durations, then expect me to choose the minimum. For example, if you include an FAT but do not state the duration as five days then I might assume three days is all that is required.

> The assumptions and clarifications section of my proposal would call out areas such as these as the basis of my bid. If I believe you left out something significant, I include an assumption stating that it is not required. The assumptions and clarifications section is my way of communicating to you, in very clear terms, what my proposal includes and excludes as well as to what degree I included various activities. When service providers are in a competitive bid situation, they have no choice but to live by the mantra "Bid the Paper" if they want to win projects.

There are some key items that should be included in the EPC service bid package to improve your chances of getting bids that are more easily evaluated and compared. The foundation of the EPC services bid package is the FEL scope document. There will likely be areas that require additional engineering depending on the granularity of your FEL. There is nothing wrong with this but provide as much detail and direction as you can as to the expected work in these engineering areas so that bidders have a reasonable bid basis.

A reference drawing package should also be included in the bid package. This reference drawing package includes all relevant drawings to the degree they are complete. For example, some of these drawing may only be "typicals" or may be overviews as previously discussed, but they should still be included. Other drawings that help orientate the bidders to the site, such as plot plan drawings, are also good to include.

Available design guide documents should also be included in the services bid package. HMI design guides, marshalling panel design guides, and alarm philosophy documents are three of the most common. If the development of any of these are being rolled into the project and are not available, you still need to provide an estimate basis for the bidders. This estimate basis may be something as simple as a description of what you expect to be included in the guide, or it may be a more detailed explanation. Keep in mind that the more details you provide, the more accurate the bidder can estimate the engineering effort associated with creating the guides.

The bid package should also contain reference work procedures or processes. This would include documents such as work permit processes (e.g., hot work requirements) and any applicable operational or maintenance procedures, which impact field engineering or field construction activities.

A standard estimate or bid tab form that outlines how you want the bid breakdown organized is also helpful. However, keep in mind that every EPC services bidder organizes their bids differently so you will need to clearly define the meaning of each of the fields. I have seen numerous bid packages that included an estimate form with no definition of any of the fields. As a result, the end user was frustrated when they received bids that deviated greatly on individual line items because bidders interpreted the fields differently. An example estimate bid tab for EPC services on a migration project is shown in Figure 3.1 below.

Billing Code	Description	Quantity	Unit of Measure	Direct Manhours	Direct Labor Pricing	Direct Material Pricing	Total Direct Field Cost
100	**Project Management**						
101	General						
102	Meetings						
103	Project Reporting						
200	**Instrumentation Design and Engineering**						
300	**Electrical Design and Engineering**						
400	**Civil-Mechanical-Building Engineering**						
500	Control Systems						
501	Design and Engineering						
502	Configuration						
503	Factory Acceptance Testing						
504	Site Acceptance Testing						
600	**Training Support**						
700	**Cutover**						
800	**Commissioning Support**						
900	**Field Construction**						
901	Construction Management						
902	Mobilization/ Demobilization						
903	Civil						
904	Instrument						
905	Electrical						
906	Control Systems						
907	Demolition						
1000	Contingency						
	Total Lump Sum EPC Service Bid						

Breakdown to an appropriate granularity that provides you with your desired bid resolution

Always include a contingency line item

Figure 3.1. Sample EPC services bid tab spreadsheet.

The bid tab spreadsheet in Figure 3.1 does not include all fields or the appropriate resolutions and is not suggested as your bid tab spreadsheet. It is intended to highlight several concepts. First, that there can be different resolutions or granularities for varying disciplines. Some of you will not likely have a need for detailed breakdown, while for others you may want to see specific subtask included. Second, you always want to include a line item that encourages bidders to state a contingency amount. They will have it in their bid and this will help you determine if they are breaking down their bid accurately. If one bidder has a contingency that is substantially lower than the other bidders, then they likely have contingency included in other budget areas. Finally, the column headers in the example provide a good starting point for building out your bid spreadsheet.

Bid instructions should be included in all bid packages. Bid instructions outline all documents that you expect to receive from the bidder including bid acknowledgements, estimates, preliminary schedules, assumptions and clarifications, etc. The bid instructions also define deadlines as well as the process for requesting additional information or asking clarifying questions. Finally, the bid instructions identify what should be included in the bid and how it should be organized if you have not provided a standard estimate form. If a standard estimate form has been included, this is a good place to define the fields of the form.

As part of the bid process, many companies have a walk-through where bidders get to see the facility, become more familiar with the existing control system equipment layout, etc. I have attended many bid walk-through meetings and their effectiveness is directly related to the engagement level of the end user. Ultimately, the goal of these walk-through meetings should be to provide all parties as much information as possible so that the bids are accurate and comparable in scope. When the end user leads the walk-through, knows the details of the project well, and engages all bidders in the process, it can help bidders with scope clarity and the identification of potential risk areas. For example, if there is a cabinet with a rat's nest wiring, the bidder may put a 10% adder into the design and engineering work for that particular cabinet. When the end user is proactively pointing out these areas to bidders, they will get back more accurate bids for the actual work.

Unfortunately, many bidders do not take advantage of the opportunity to address areas of confusion or concern during the walk-through process. In most cases, this is because they are afraid that if they ask questions they will give something away about their approach or strategy. These concerns are largely unwarranted and only damage the EPC services bidder's ability to fully understand all aspects of the project leading to assumptions that may or may not be accurate. Approach the development of your EPC services bid package focused on ensuring that bidders will have all of the information they need to include everything you expect in their bids.

Bid Evaluation and Project Award

The bid evaluation process for EPC service providers should mimic the process used for the evaluation of control system bids. The process is to define and prioritize decision criteria, and then rank the bidders based on those weighted criteria. Criteria for EPC services bids include everything from familiarity with both the old and new control systems to capabilities to integrate with third-party systems such as historians.

There are a number of common bid practices that you want to be aware of as you evaluate bids. EPC services bidders tend to look for areas to reduce the costs and hide many of these in subtle assumption statements. For this reason, it is important to thoroughly examine the assumptions and clarifications of all bidders. The assumptions are where most of the bid deviations occur. Many bidders use assumptions and clarifications correctly and are diligent about trying to establish an accurate view of their bid basis. However, there are other bidders who will use the assumptions and clarifications as a way to try to omit items from their bid. If you find any large discrepancies in the assumptions, clarify the particular assumption point and ask for all bidders to modify their bids based on your clarification.

If you fail to fully define your requirements in any scope area, then it will either be left out of bids or will be included to meet minimum requirements. For example, if you say project management reporting is required but don't spell out the frequency and type, then expect bidders to take a minimalistic approach. I re-emphasize that if you want something included in your bid from EPC service providers be as specific as possible quantifying where you can so that you eliminate the opportunity for inconsistent interpretations and varying assumptions.

SUMMARY

Bid specifications and the vendor selection process for the control system as well as the EPC services can often be overlooked as an integral part of the migration project. The reality is that short-changing these areas can get your project offtrack early. The amount of time invested in these documents will directly impact the quality of the bids that you receive.

The objective of the bid process is to achieve accurate and comparable bids. To effectively meet this objective, you must provide a clear and detailed scope as the bid basis. Aligning expectations in key areas during the bid process typically reduces frustrations of both the end user company and the vendors while decreasing change orders during project execution.

The evaluation of vendors for both the control system and the EPC services can be an emotional process. There are often loyalties that create conflict. Using the decision criteria process outlined in this chapter can help you better understand the priorities for your vendor selection and remove a significant part of the subjectivity to reach a logical rather than emotional decision.

Three Key Takeaways

- Be as specific and complete as possible in defining your requirements to enable more accurate bids and easier vendor comparisons.
- Use the decision criteria matrix process to define and prioritize your needs and reduce the emotional aspects of vendor selection.
- Separate the control system bids from the EPC services bids to gain a better understanding of project cost and more accurately compare bids.

4

Scope, Schedule, and Budget

Complete and detailed scope, schedule, and budget documents are fundamental to good project management. At the most basic level, the project scope of work should focus on maintaining consistency with the goals established when the project was initiated. Good tools to reference as you build your project scope include project charter, project justification, and AFE documents as well as any previous scope of work versions such as those that might have been produced in early FEL stages.

Thorough scope of work documents on control system migration projects not only define what needs to be done, but also outline how the work should be accomplished. They also include relevant details such as quantities and complexities associated with the work. For instance, if you have graphics that must be configured on your project, a good scope of work would identify that ten process graphics of medium complexity are required. It would also include a definition of what represents a medium complexity graphic based on factors like the density of static equipment and piping on the graphic, the number of dynamic elements, required customization of graphic symbols or faceplates, etc. Including a higher resolution of work details such as these in your scope will make it much easier to develop an accurate budget estimate and schedule during the definition and detailed engineering phases of the project. In addition, better scope granularity makes scope changes during project execution more obvious and simplifies progress tracking and reporting.

The scope, schedule, and budget are directly related to one another and must be consistent using the same terminology and addressing similar levels of detail. This will simplify project management progress monitoring and reporting when the project gets underway. In addition, each of these documents serves as a good crosscheck and validation for others. The example below shows how validating consistency between the scope, schedule, and budget can help identify potential issues in the project so that they can be proactively managed and addressed through project planning.

Example

Your project scope includes configuring 50 process graphics. You have estimated that each of these graphics takes four hours to build and then must be reviewed by operations. Your budget includes a junior graphics designer at $70 per hour building these graphics with a $14,000 total budget for this line item. In your schedule, you have included this line item at three weeks. When you compare the budget and schedule you recognize a discrepancy.

Your estimate assumes 200 hours of work (i.e. 50 graphics × 4 hours per graphic). If a single junior graphics designer is working on the project and is not permitted to work overtime, then the schedule would require a minimum of five weeks (i.e. 200 hours ÷ 40 hours per week). It should also be noted that the schedule does not take into account the operations review cycle. Any rework associated with comments from the review cycle should also be factored into both the schedule and budget.

There are a few options to address the schedule and budget discrepancies. The schedule line item of three weeks can be updated to reflect the longer duration of five weeks if a single resource is assigned to the activity. The second option is to add a resource so that the original three-week schedule can be met. If this is done and the second resource is at the same bill rate, then this will have no budget impact. If the resource is at a higher bill rate, it could impact the budget for this activity. Regardless of what adjustments are made, the operations review cycle will also need to be integrated into the schedule and budget.

If the scope, schedule, or budget is misaligned, such as was the case in this example, then assuming a constant scope you will have issues meeting either your schedule or budget. The relationship between scope, schedule, and budget for any project is fixed. This is commonly represented in the project management triangle as shown in Figure 4.1, which is also often referred to as the triple constraint triangle. While a simplistic depiction, understanding this basic triangular relationship is fundamental to good project management.

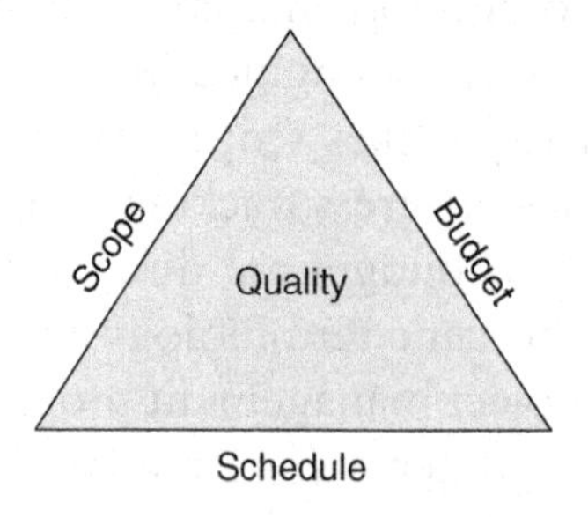

Figure 4.1. Basic project management triangle.

The fixed relationship between scope, schedule, and budget is intuitive but can sometimes be difficult to accept when put into practice. End user companies often request changes to scope, schedule, or budget from third-party service providers not fully understanding the correlation and with the expectation that it will not result in change orders for the project. Expecting that the cost or schedule impact of changes will somehow be absorbed into the existing project is unrealistic. Changing any of these three elements will impact the other elements. If you reduce the budget, then expect the scope to decrease or the schedule to be extended. If you compress the schedule, expect the budget to increase or the scope to be reduced. The example below reinforces the fixed relationship between scope, schedule, and budget and shows how adjustments to one of the elements can affect the others.

Example

Your company has planned a 20-day plant turnaround during the summer. Due to this window of opportunity, your project team decides to perform a cold cutover during this outage. As company planning for the turnaround progresses, the duration is compressed from 20 days to 16 days. How does this schedule change impact the scope or budget of your project?

We will assume your project scope stays fixed for the turnaround. To accomplish this same scope in less time will require additional resources or significant overtime. Both of these resource options will increase the budget. If the budget is fixed and cannot be increased, then a second option is to reduce the scope of work for the cutover. How can this be done? You may potentially be able to shift some of the work to a hot cutover period prior to the outage so that you have a more manageable scope during the turnaround. This option may also have some impact on the budget if additional resources are required. However, it may minimize the impact on cold cutover scope execution per the original staffing plan.

While the example above emphasizes the scope, schedule, and budget, do not forget the relationship to quality that each of the elements has as well. Compressing schedule, reducing budget, or expanding scope can all impact the resourcing plan and have unintended consequences on work quality. This is why any changes during your project should be well thought out and their impact completely understood before proceeding.

The scope, schedule, and budget can also be valuable project management tools for validating that project staffing plans are reasonable. For each major task, you should be able to evaluate and compare the planned number of resources, proposed billing rates, and expected time allocations against budget and schedule documents for consistency. Any inconsistencies are a warning that your staffing plan may be a project constraint, or itself be constrained by your scope, schedule, or budget.

In the earlier chapter on FEL deliverables, we discussed scope, schedule, and budget in general terms. In this chapter, we will examine the details of how to organize an effective scope of work. We will review by discipline the common work scopes and deliverables involved in control system migration projects. We also outline how to build effective budgets and schedules with an appropriate resolution level for effective project management.

SCOPE

The completeness of the scope is essential for the project manager to effectively monitor and evaluate progress during the control system migration project. The scope of work document should at a minimum address all engineering, procurement, and construction work associated with all functional discipline areas required for the project. Any additional information such as the quantities or work complexity definitions mentioned earlier that help build boundaries around the scope and frame the work to be accomplished should also be included.

Depending on the granularity of the scope developed during the FEL, it may or may not be adequate for the purposes of project management. It is common for some engineering areas to need additional definition and scoping during the migration project itself. For the purposes of project management, you can initially include placeholders in your scope, schedule, and budget until the details are finalized, but it is important to prioritize any nebulous scope areas and finalize them as early in the project execution phase as possible. The way the FEL scope is organized may be an additional area that requires some adjustments to make the scope more meaningful for project management purposes as the project moves forward.

Overall Organization and Approach

For project managers, it is valuable to organize the scope, schedule, and budget in a similar way. This makes it much easier to ensure consistency among the documents and also simplifies the process of answering questions and clarifying details throughout the project. There are numerous ways to organize a scope of work document. None are necessarily wrong but certain approaches can sometimes create challenges for project managers. In Table 4.1 below, I have outlined several common ways that I have seen control system migration scope of work documents organized along with some comments on each.

Table 4.1. Common approaches to organizing scope of work documents

Scope of Work Organizational Style	Comments
Physical location or area	This approach organizes work activities by physical work location or area. This may be as simple as work scope in the field versus in the control room. Alternately, it may be work scope in various operating units and areas within the facility. This approach is typically well liked by the field construction team. It can also be helpful on larger projects with a number of units or areas. However, this can be a difficult approach for the project manager because all work discipline categories must be tracked by area or location causing substantial repetition of subtasks under each area.
EPC category	This approach groups all work for engineering in one category, all procurement in another, and all construction in another. This approach can work well for project managers because it can simplify resource planning and enable independent schedule creation for each work group. However, this approach creates challenges when change orders are required. A single change order may impact engineering, procurement, and construction work groups. The project manager must divide the change order among each function creating numerous new entries into the budget and schedule for a single change order.
Functional discipline	This approach organizes work activities by functional discipline, such as instrumentation, electrical, and controls. This can be an effective way for the project manager to organize the budget and schedule. The downside is that it lumps engineering, procurement, and construction activities for each discipline together meaning it may be more difficult for individual workgroups to isolate their particular work tasks.

I generally prefer organizing the scope of work, schedule, and budget by functional discipline. This is a personal preference and the other strategies can be just as effective. For the remainder of this chapter, I will be using the functional discipline organizational approach as my basis. The following discipline areas will be covered as key sections of the control system migration scope:

- Instrumentation
- Electrical
- Controls
- Civil-Mechanical-Building
- Communications and Integration

- Testing
- Training and Documentation
- Cutover

The scope of work associated with each of these areas is variable between projects. For example, relative to the instrumentation scope, not everyone will have a field junction box so the instrument may be wired directly to the marshalling location meaning that activities related to field junction box design and engineering are not applicable. However, I have included some checklists to provide guidance in considering which common work activities are part of your scope. The included checklists identify major activities only and are not intended to cover every possible activity or each subtask that comprises an activity. While the checklists provided are good starting points, every project team should develop their own checklists for given discipline areas to ensure no important project tasks have been omitted from your scope, schedule, and budget documents.

There are some common responsibilities applicable to many and in some cases all of these functional discipline areas. I have not specifically included them in the checklists to avoid repetition. As an example, quality assurance should be a part of every functional discipline area. When work tasks or deliverables are completed a quality assurance process needs to be in place to review the work. Demolition activities may also be a part of multiple functional discipline areas and impact the design, engineering, and construction work scopes.

Instrumentation

There are many facets to the instrumentation scope of a control system migration project. The design and engineering scope for instrumentation begins with updating or generating drawings such as instrument loop drawings, wiring and cable tray drawings, and termination schedules. The instrument design and engineering function typically also owns the I/O index or instrument database, which is a critical documentation component that must be kept up to date throughout the project and shared with the controls engineering function and the field construction team.

There are mature software tools (e.g., Intergraph® SmartPlant® Instrumentation) that greatly simplify the process of building and maintaining an I/O index. Many of these software packages can also support automated drawing generation. If you are not employing these tools, you might consider whether they can be used to save time and reduce budget on your project. Particularly on projects with a large number of I/O and loops, the time-saving benefits can help pay for the software tools.

If your migration project includes re-instrumentation or added instrumentation, then installation details such as calibration instructions, instrument power details, instrument mounting design, and instrument signal grounding practices are required. The design and engineering associated with the routing and termination of any new or revised signal wiring to the control system is another critical component of the instrumentation scope. The major segments of the instrument wiring design scope include:

- The field component, which encompasses the field instrument and junction box wiring and terminations with associated cable tray and conduit routing details.
- The marshalling component, which includes marshalling cabinet installation details, marshalling panel layouts, and all associated wiring, termination, tagging, conduit, and cable tray details.

Many projects include the final signal wiring segment from the marshalling cabinet to the I/O cabinet under instrumentation responsibilities as well. I recommend managing this scope under the control design and engineering scope because the marshalling panel is a natural dividing line where you transition from the field instrument to the control system equipment. From a practical standpoint, the same team may be doing the work, but grouping this under controls design and engineering makes tracking it in the budget and schedule more logical. It can also help promote communication between instrumentation and controls design and engineering resources.

If the existing instruments are not being replaced or upgraded, the purchasing responsibilities associated with the instrumentation scope will be limited to marshalling cabinets and commodity materials such as conduit, cable tray, cables, and labels. The specification and purchasing of instruments and instrument stands will become part of the scope if instruments are being replaced or added as part of your migration project.

The instrumentation-related construction scope includes installing marshalling cabinets, instrument stands, instruments, and any additional field junction boxes. The construction team is also responsible for installing cable tray, conduit, and cable for the runs from the instrument to any field junction boxes and then to the marshalling cabinets. Labeling of all wiring and terminals as well as the final terminations at the instrument, field junction boxes, and marshalling panel are all included in the instrument construction scope as well.

A general checklist for the instrumentation scope is provided in Table 4.2 below.

Table 4.2. Instrumentation scope checklist

Description	Category	Required
New or replacement instrument specification and installation design details	Design/engineering	Yes ☐ No ☐
Instrument wiring, conduit, and cable tray drawings from field device to field junction boxes	Design/engineering	Yes ☐ No ☐
Signal wiring, conduit, and cable tray drawings from field junction boxes to control system marshalling	Design/engineering	Yes ☐ No ☐
I/O index development and maintenance	Design/engineering	Yes ☐ No ☐
Marshalling panel design layout	Design/engineering	Yes ☐ No ☐
Loop sheet drawings	Design/engineering	Yes ☐ No ☐
Field junction box termination schedules	Design/engineering	Yes ☐ No ☐
Control system marshalling termination schedules	Design/engineering	Yes ☐ No ☐
Voltage drop calculations	Design/engineering	Yes ☐ No ☐
Purchase of instruments, instrument stands and commodity I&E items (cable tray, conduit, terminals, etc.)	Purchasing	Yes ☐ No ☐
Install marshalling cabinets and panels	Construction	Yes ☐ No ☐
Install new-field junction boxes	Construction	Yes ☐ No ☐
Install field instruments	Construction	Yes ☐ No ☐
Install cable tray, conduit, and cable from field instrument to field junction box	Construction	Yes ☐ No ☐
Install cable tray, conduit, and cable from field junction box to control system marshalling	Construction	Yes ☐ No ☐
Attach wiring labels and complete terminations at field device, junction box, and marshalling panels	Construction	Yes ☐ No ☐

Electrical

The electrical scope is generally not a large component of control system migration projects. If there are significant electrical infrastructure changes that need to be made to support the control system migration project, these changes are usually handled as separate projects. The electrical design and engineering scope for a control system migration should include a load study to ensure that the power requirements of the new system will not exceed the existing infrastructure's capacity. These load studies are often focused on the transition period when much of the existing control system remains in place, while components of the new system are added resulting in a maximum electrical load scenario.

The electrical scope design and engineering also includes developing or revising power panel termination schedules, power panel installation details, and uninterruptible power supply (UPS) installation or connection details. Electrical cable tray, conduit, and cable routing plans, and terminations schedules must also be designed for the new control system. Grounding grid evaluations and electrical grounding instructions are also a part of the electrical scope.

In recent years, the electrical infrastructure has become more tightly integrated with control systems. For example, some vendors today offer intelligent electrical motor management and control devices that interface to the control system. Many end user companies are recognizing the benefits of this capability and are increasingly adopting the use of these devices. When these devices are a part of your project, configuration of the interface is required and most often supported by combinations of the I&E and control system design and engineering teams.

The electrical purchasing scope of most control system migration projects includes commodity items such as cables, connectors, conduit, cable tray, breakers, and power panels. The construction scope covers installing any new power panels, running power cables, and any associated tray or conduit to final locations for all control system equipment. Labeling and termination of all cables is also a construction scope item. Temporary or transition construction activities are also sometimes required and might include installation of temporary power or routing of power to interim control system equipment locations.

A general electrical scope checklist is shown in Table 4.3 below.

Table 4.3. Electrical scope checklist

Description	Category	Required
Power panel layout and termination schedules	Design/engineering	Yes ☐ No ☐
Power load study calculations	Design/engineering	Yes ☐ No ☐
Control system power supply and UPS installation details	Design/engineering	Yes ☐ No ☐
Grounding assessment and equipment grounding instructions	Design/engineering	Yes ☐ No ☐

(*Continued*)

Table 4.3. (*Continued*)

Description	Category	Required
Interim/temporary power plan	Design/engineering	Yes ☐ No ☐
Power wiring, conduit, and cable tray drawings to all cabinets and workstations	Design/engineering	Yes ☐ No ☐
Configuration of interfaces for intelligent motor management and control devices	Design/engineering	Yes ☐ No ☐
Purchase of power supplies, power panels, and commodity electrical items (breakers, cable tray, conduit, terminals, etc.)	Purchasing	Yes ☐ No ☐
Rental agreement or purchase for temporary power needs	Purchasing	Yes ☐ No ☐
Install all required power panels, cable tray, conduit and cable	Construction	Yes ☐ No ☐
Install temporary power circuitry and interim power cabling	Construction	Yes ☐ No ☐
Install power supplies and UPS system	Construction	Yes ☐ No ☐

Controls

The controls design and engineering scope is usually the core component of the migration project. This function includes designing the final architecture of the control system including physical location drawings for control system equipment. The controls design and engineering scope also includes control system I/O addressing, I/O cabinet installation details, I/O cabinet panel layouts, I/O termination schedules, and I/O tagging details. The wiring interconnection details between the marshalling panel and the I/O cabinet are also a part of this scope.

The system configuration scope includes executing the basic configuration activities for control system hardware and I/O points. System administration responsibilities include the initial configuration of passwords, system security settings, and alarm conventions (e.g., priorities). The system configuration function also includes documenting and configuring all loops and other advanced logic such as interlocks, sequencing, and complex logic algorithms. Graphics also represent a major component of the controls configuration scope.

The control system is the primary purchasing item for a migration project. There are times that purchasing of the control system is a part of the EPC services responsibility, but in many cases, end users choose to purchase the system directly. Control system vendors prefer the direct purchase from the end user because it enables them to better track how, when, and where their systems are installed. In either case, the control system engineering responsibility includes developing a complete bill of materials for purchasing activities. Controls purchasing responsibilities also include commodity items such as terminals, cable, and wiring duct related to the control system installation.

The field construction scope associated with the control system involves the installation of all components of the control system from controllers to operator workstations. This includes installing all interconnectivity wiring for the system to function properly. One of the biggest challenges of control system field construction is working around the existing system in an active operating environment. The transition periods when the old system is still being used to operate the unit and the new system is partially installed can present several issues such as limited physical space in which to work. Another common challenge during this time is minimizing disruption to the normal operating tasks in the control room.

Legacy control systems were built into specially designed furniture. Modern control systems can be integrated with most commonly available desktop surfaces. As a result, when replacing older control systems, new control room furniture is often included in the project scope. The detailed layout design as well as specification, purchasing, and installation of the control room furniture have become an integral part of the control scope for many migration projects.

A general controls scope checklist is provided in Table 4.4 below.

Table 4.4. Controls scope checklist

Description	Category	Required
Final control system layout architecture drawing	Design/engineering	Yes ☐ No ☐
Control system addressing assignments	Design/engineering	Yes ☐ No ☐
I/O cabinet layout drawings	Design/engineering	Yes ☐ No ☐
Control room layout drawings	Design/engineering	Yes ☐ No ☐
Temporary control room drawings	Design/engineering	Yes ☐ No ☐
Wiring, conduit, and cable tray drawings from marshalling cabinets to I/O cabinets	Design/engineering	Yes ☐ No ☐
Control system equipment installation details (controllers, I/O cards, workstations, etc.)	Design/engineering	Yes ☐ No ☐
Control room furniture layout drawings and installation details	Design/engineering	Yes ☐ No ☐

(*Continued*)

Table 4.4. (*Continued*)

Description	Category	Required
Purchase of control system equipment and commodity items (cable tray, conduit, terminals, wire labels, etc.)	Purchasing	Yes ☐ No ☐
Purchase of control room furniture	Purchasing	Yes ☐ No ☐
Configuration of I/O points	System configuration	Yes ☐ No ☐
Configuration of logic and loops	System configuration	Yes ☐ No ☐
System hardware configuration	System configuration	Yes ☐ No ☐
Configuration of graphics	System configuration	Yes ☐ No ☐
System administration configuration (e.g., Security, password protection, alarm priorities, etc.)	System configuration	Yes ☐ No ☐
Install I/O cabinets	Construction	Yes ☐ No ☐
Install controllers and I/O cards in I/O cabinets	Construction	Yes ☐ No ☐
Install control room furniture	Construction	Yes ☐ No ☐
Install operator workstations	Construction	Yes ☐ No ☐
Install all required cable tray, conduit and cable between marshalling and I/O cabinets	Construction	Yes ☐ No ☐
Terminate power cables for all control system equipment	Construction	Yes ☐ No ☐
Label and terminate all control system I/O wiring	Construction	Yes ☐ No ☐

Civil-Mechanical-Building

There may be minimal or even no civil-mechanical-building discipline scope involved in your control system migration project. However, many control system projects do involve this component to varying degrees. If your project requires installation of a new remote instrument enclosure, building of a new control room, or revisions to an existing control or marshalling room building, then you will have a civil-mechanical-building component. Specifically, if any new buildings are to be installed on site, then a facility siting survey will need to be completed or an existing one revised ensuring that the building is located in a safe area.

Outside of facility siting efforts, an example of a common civil-mechanical-building discipline work task is designing and constructing slabs for new buildings. Whether you are using an existing building or installing a new RIE building if additional cable, conduit and tray entries are required to the building from outside then this design should also be completed by the civil-mechanical-building discipline. In addition, any structural supports that are required for cable trays, marshalling, and I/O cabinets, etc., may require civil design and engineering. This functional discipline is not only the civil and mechanical aspects but can also include other building design and engineering consideration areas, such as HVAC capacity and heat load dissipation.

A general checklist for a migration project civil-mechanical-building scope is provided in Table 4.5 below.

Table 4.5. Civil-mechanical-building scope checklist

Description	Category	Required
Development of or modification to facility siting survey	Design/engineering	Yes ☐ No ☐
Revisions to control room building drawings	Design/engineering	Yes ☐ No ☐
Revisions to marshalling room building drawings	Design/engineering	Yes ☐ No ☐
New or revised remote instrument enclosure drawings	Design/engineering	Yes ☐ No ☐
Revisions to plot plans, underground drawings, etc. (required for new buildings)	Design/engineering	Yes ☐ No ☐
Design of supports for cable trays and cabinets	Design/engineering	Yes ☐ No ☐
New or revised drawings for cable, conduit and tray entries into buildings	Design/engineering	Yes ☐ No ☐
HVAC and heat load studies	Design/engineering	Yes ☐ No ☐
Purchase of RIE buildings, construction materials, cement for foundations, etc.	Purchasing	Yes ☐ No ☐
Pouring slabs for buildings	Construction	Yes ☐ No ☐
Installation of new RIE buildings and associated equipment (e.g., HVAC)	Construction	Yes ☐ No ☐
Demolition and reconstruction as required for control room and marshalling room buildings	Construction	Yes ☐ No ☐
Punch holes in buildings as needed for cables, conduit and tray entries	Construction	Yes ☐ No ☐

Communications and Integration

The communications and integration scope is sometimes grouped with the control system scope. This is fine if you have minimal third-party connectivity requirements, but in general I recommend keeping this as a separate function. The design and engineering responsibilities related to the communications and integration scope begin with the development of a master list of all third-party systems and applications with which the new control system must interact, the installed hardware and software versions and the type of interface (e.g., OPC, Modbus, etc.) required. A list of third-party devices, systems, and applications commonly connected to the control system is provided in Table 4.6 below.

Table 4.6. List of common third-party connections to control systems

Programmable logic controllers (PLCs)	Safety instrumented systems (SIS)	Data historians or manufacturing execution systems (MES)
Lab information management systems (LIMS)	Process analyzers	Vibration monitoring systems
Advanced process control (APC) applications	Batch management applications	Inferential modeling software
Alarm management systems	Enterprise resource planning (ERP) solutions	Product scheduling software

The design and engineering scope includes fully documenting any cabling, switches, hardware, and software interfaces required for the communications in network diagrams or other representative drawings. Third-party connections to the legacy control system are often not well documented. As a result, it is very important that these connections are thoroughly investigated and researched to identify any unique requirements or application customizations. The design and engineering function is usually also responsible for gathering documentation on the setup and configuration of the communications in support of project installation activities. The control engineering function configures these devices and verifies they are communicating properly. Formal verification testing of third-party system and application connectivity is also often integrated into the Factory Acceptance Test (FAT) plan.

The degree to which the control system communicates or exchanges information with these third-party products varies among companies. Completing a throughput analysis and validating that communications bandwidth will not be a problem is essential. Another critical design consideration today is the need to ensure the security of these communications to maintain the integrity of the connection.

There are additional communications devices that may not be directly interfaced to the control system but are still an important part of the work-station environment design considerations. These include peripheral devices such as radios, cameras, phones, standalone PCs, and hardwired shutdown switches. The design of all of these must be considered as part of the overall control room layout. The installation or relocation of these devices is part of the construction scope. While generally straightforward, shifting these devices from the old control system workstation environment to the new control system workstation environment requires planning and coordination with the operations, maintenance, and IT departments.

In Table 4.7 below is a checklist of communications and integration scope items.

Table 4.7. Communications and integration scope checklist

Description	Category	Required
Develop and maintain master list of required third-party interfaces and connectivity	Design/engineering	Yes ☐ No ☐
Develop interface network diagram	Design/engineering	Yes ☐ No ☐
Third-party hardware, software and interface installation details	Design/engineering	Yes ☐ No ☐
Configuration of required interfaces	Design/engineering	Yes ☐ No ☐
Verification testing of connectivity and communications	Design/engineering	Yes ☐ No ☐
Addition of communication devices (e.g., phones, cameras, etc.) To control room layout drawings	Design/engineering	Yes ☐ No ☐
Purchase of required network switches, hardware, software, or interfaces	Purchasing	Yes ☐ No ☐
Installation of required hardware in workstations	Construction	Yes ☐ No ☐

Testing

As mentioned and defined in the earlier FEL deliverables list, the two tests included in most control system migration projects are the FAT and the Site Acceptance Test (SAT). The FAT and SAT detailed test plans are generally developed later in the project rather than at the FEL stage as part of the design and engineering team work responsibilities. Documenting these test plans and keeping signed records of the completed test results are required by some regulatory agencies depending on your global location, the nature of your control system (e.g., safety instrumented system), and your industry.

A complete FAT plan should first establish that all hardware and software components of the system are working properly. This involves checking all major components of the control system including workstations, controllers, and I/O. The degree of I/O checkout can vary and in some instances may require I/O card point-to-point verifications while in others spot checking I/O card functionality suffices. If the system has redundancy, the failover performance is demonstrated as part of the FAT. The proof of connectivity to third-party systems or applications is also often included in the FAT. This can sometimes be a challenge as it requires the third-party system, software, or application to be available at the FAT location. However, proving this connectivity before site installation is important because communications issues with third-party solutions can be difficult and time-consuming to troubleshoot.

Thorough FAT plans also require testing of configuration and graphics. Simulation packages, either internal to the control system or external via third-party software, can be used to simulate I/O signals. Signal drivers at the I/O card can also be used to accomplish the same thing, although this is a much more time-consuming approach. The configuration is usually spot checked with a representation of various types of signals within each controller to validate that the configuration is correct. For instance, you want to verify that controllers drive valves to the appropriate fail open or fail closed position. Other checks may include verifying tuning settings and testing controller response to out-of-range instrument signals.

Logic configuration is also checked during an exhaustive FAT. The testing of logic is usually limited to spot checks because detailed testing of all logic requires so much time. Spot checks are usually done for each of the different logic categories to verify that the general logic algorithms perform as intended. The final part of an FAT is verifying graphics functionality. This includes using faceplates for control actions and validating that alarm information is displayed correctly and logged accurately. A spot check of graphics for properly functioning dynamic elements might also be done as part of the FAT.

To effectively manage an FAT, a process needs to be established for documenting issues or errors, making corrections, and retesting. This process varies greatly but having a complete punch list of all items that are identified during the FAT as needing modification is critical. Whether the individual issues are corrected as part of the FAT or handled on an individual basis between the FAT and SAT, the process needs to determine how corrected issues will be rechecked and who is responsible for final approval.

The SAT plan is also a common scope deliverable. The SAT includes onsite verification of the system confirming power, grounding, and signal integrity prior to cutover activities. The degree of detail within an SAT depends on the cutover approach and the availability of physical space for staging among other factors. The actual SAT execution is done on site and requires project management involvement for scheduling and logistical planning.

The potential procurement activities associated with testing are leasing or purchasing hardware, software, or other equipment needed for testing activities. If staging area space is not otherwise available, procurement may also be involved in leasing temporary space. Construction involvement is often not required for testing but may be utilized to setup a temporary staging area for the FAT. The construction team installs the equipment at the worksite as needed for the SAT as well.

A general checklist for testing scope items is provided in Table 4.8 below.

Table 4.8. Testing scope checklist

Description	Category	Required
Develop factory acceptance test (FAT) plan	Design/engineering	Yes ☐ No ☐
Execute FAT plan	Design/engineering	Yes ☐ No ☐
Resolve FAT punch list items	Design/engineering	Yes ☐ No ☐
Develop Site Acceptance Test (SAT) Plan	Design/engineering	Yes ☐ No ☐
Execute SAT Plan	Design/engineering	Yes ☐ No ☐
Resolve SAT punch list items	Design/engineering	Yes ☐ No ☐
Lease temporary facilities for FAT	Purchasing	Yes ☐ No ☐
Lease or purchase hardware, software and miscellaneous materials required for testing	Purchasing	Yes ☐ No ☐
Install or set up temporary equipment in staging area for FAT	Construction	Yes ☐ No ☐
Install equipment for SAT	Construction	Yes ☐ No ☐

Training and Documentation

It is far too common for training and documentation to not be prioritized until well into the project. The planning for these two areas should actually be done early in the project. They are critical aspects of a successful control system migration project. A later chapter will discuss training requirements for control system migration projects in detail, so here I will only outline the scope deliverables associated with training.

The engineering and design team is responsible for establishing a comprehensive training plan. This plan development is typically done in conjunction with the training department and also involves the individual departments such as operations and maintenance to the degree needed. A comprehensive training plan should outline the way that the appropriate knowledge of the new control system will be conveyed to all of those who interact with the control system to perform their job responsibilities. The procurement scope for training is purchasing training services from the vendor and in some cases renting temporary training space. Construction generally has no scope associated with training activities.

Engineering and design is responsible for collecting and assembling all final project documentation in conjunction with the project manager. This can be an extensive list of items that vary by project, but should include drawings, specifications, equipment manuals, I/O list, and other documents important for capturing all work done on the project and assisting in maintenance of the installed system moving forward. Procurement is responsible for providing all purchase orders associated with the project to the engineering and design team to include in the final project documentation. Construction also plays a vital role in project documentation. The construction team is responsible for communicating markups and redlined drawings for any deviations of the installation from the original design to the team. The design team will then use these markups to create the final as-built drawings.

Training and documentation scope of work items are listed in Table 4.9 below.

Table 4.9. Training and documentation scope checklist

Description	Category	Required
Develop operations training plan	Design/engineering	Yes ☐ No ☐
Develop engineering training plan	Design/engineering	Yes ☐ No ☐
Develop maintenance training plan	Design/engineering	Yes ☐ No ☐
Operations training	Design/engineering	Yes ☐ No ☐
Engineering training	Design/engineering	Yes ☐ No ☐
Maintenance training	Design/engineering	Yes ☐ No ☐
Lease temporary facilities for training	Purchasing	Yes ☐ No ☐
Procure training services and equipment as required	Purchasing	Yes ☐ No ☐

(*Continued*)

Table 4.9. (*Continued*)

Description	Category	Required
Provide all initial project design and engineering drawing and documents to field construction teams	Design/engineering	Yes ☐ No ☐
Convert redlined drawings and markups to final as-built deliverables	Design/engineering	Yes ☐ No ☐
Collect and assemble final project documentation	Design/engineering	Yes ☐ No ☐
Provide all project purchase orders to design and engineering	Purchasing	Yes ☐ No ☐
Markup or redline project drawings and documentation when installations deviate from original design	Construction	Yes ☐ No ☐

Cutover

Successful control system migration cutovers require a well-conceived strategy, detailed upfront planning, and frequent communication during the execution phase. A later chapter will be dedicated to discussing best practices associated with the management and execution cutovers. In this section, we review the work scope associated with cutovers.

The engineering and design team is responsible for investigating the details and developing the documentation related to the cutover. This begins with a thorough assessment of cutover options and finalizing the detailed cutover plan. Loop folders and segmented documentation packages are developed and used to help organize information around the cutover schedule. For instance, you may be managing the cutover by I/O cabinet so the documentation is segmented into cutover packages for each I/O cabinet. Typically, an engineering resource also has responsibility for managing the cutover and loop checkout activity in the control room.

There are no special purchasing responsibilities associated with the cutover other than potentially requiring rental or procurement of temporary power or equipment to help support the cutover activities. The construction team is heavily involved in the cutover activities. Note that when I say construction team, it may be the company I&E department, but ultimately refers to the resources completing field construction activities by finalizing terminations and performing final loop checkout.

In Table 4.10 below is a general checklist for cutover scope of work items.

Table 4.10. Cutover scope checklist

Description	Category	Required
Develop detailed cutover plan	Design/engineering	Yes ☐ No ☐
Create loop folders	Design/engineering	Yes ☐ No ☐
Organize and segment cutover documentation	Design/engineering	Yes ☐ No ☐
Coordinate loop checkout in control room	Design/engineering	Yes ☐ No ☐
Purchase or lease interim equipment or power as required	Purchasing	Yes ☐ No ☐
Make final terminations per cutover plan	Construction	Yes ☐ No ☐
Execute checkout from field	Construction	Yes ☐ No ☐

BUDGET

The control system migration project budget frequently has the highest visibility with management. Costs are always a priority. As a result, it is important to accurately establish the project budget and focus on managing the scope and schedule to it. Determining the right work breakdown structure is a critical part of establishing a budget that is both manageable and easy to track. If you track the work activities at too high of a level, it will be difficult to proactively identify when in-progress tasks are taking longer than budgeted. This often leads to a false sense of security and the project manager can easily fail to recognize that an activity is trending over budget. This can mean that problem areas are often not identified until it is too late to take corrective action.

Tracking work activities at too granular of a level can also be problematic. Too much detail can be difficult to manage. Highly granular budgeting can also mean resources spend additional time trying to track and report their time. In this case, the project management process requires too much time be invested by not only the project manager, but also project resources with little benefit for the additional time required. A good balance between detail and management efficiency needs to be found for effective budget tracking and reporting. The granularity needs to provide enough insight so that it is easy to identify when in-progress tasks are trending over budget, but at the same time not to a level that is cumbersome to the project team or unwieldy for the project manager. The example below demonstrates how different budget work breakdown resolutions can be used and how they correspondingly impact the ability to monitor and track spending against the budget.

Example

A project task for your migration is the configuration of three different types of advanced logic algorithms in the control system. There are five of Type A, three of Type B, and four of Type C to configure and you need to establish an effective work breakdown structure for your budget.

The first option is to lump all of these together in a single budget line item that covers advanced logic configuration. This might be an effective strategy if you are certain that there is limited variability in the amount of configuration time required for each. However, this will give you little feedback while the work activity is in progress as to whether a task is on budget or trending over.

The second option is to establish a budget line item for each logic type. In this case, you would have three line items, one for each type. This allows you to track each major type and ensure your configuration time assumptions for each are correct. For instance, you may have assumed that to configure Type A logic will take four hours each. If you see that for this line item you actually find that the first two took 16 hours, then you can see that this logic configuration is trending double the budget. You have the opportunity to make adjustments to the budget for the remaining logic configuration tasks or put a corrective action plan in place to recover the overage elsewhere in the project.

The final option is to track it at the individual subtask level. This would mean that you actually list each of the Type A, Type B, and Type C logic algorithms to be configured as individual budget line items. In this instance, you would have twelve line items for logic configuration. This gives you the highest resolution but also creates a very large budget spreadsheet. It will require the resources on the project to breakdown their specific work to a very detailed level. This strategy is typically used for areas of high risk of variability where you want to closely track progress on a daily basis but otherwise is considered too detailed.

After evaluating these options, the second option is likely the most advantageous. It tracks the logic algorithms with enough resolution so that you can see where the logic configuration is actively trending versus budget but without requiring such detail that it becomes difficult to manage.

Building a budget that is based on labor services, particularly those involving a number of parties, can be challenging in part because rates are variable based on the individuals working on any given tasks. For instance, if a junior designer is working on verifying the I/O documentation for a particular marshalling cabinet it may be at a rate of $60 per hour. If the cabinet next to it is being verified by a senior designer, the rate may be $80 per hour. Accounting for the staffing mix is one of the major challenges of effectively establishing and managing budgets on any labor services heavy project such as a control system migration.

There are several basic fields essential to a labor services budget. For labor services tasks, the two core fields that you want to include are the hourly rate of the resource doing the work and the budgeted number of hours to complete a task. When multiplied together the result is the total budget for the line item task. To improve the granular resolution and enhance the ability to make changes in the budget as work among staff resources is shifted, added, or removed, extra columns for quantities and bill rates per discipline category can also be added.

Figure 4.2 below illustrates example headers for the most basic method and then the more detailed method. In the case of the most basic method, note that the bill rate is a composite rate of several resources and must be predetermined. In the more detailed header, the calculations are embedded in the spreadsheet so that as changes are made to the bill rates or quantities the budget calculation is automatically updated.

Item Description	**Estimated Hours**	**Hourly Rate**	**Budget**		
Configure Graphics	200	$81.25	$16,250.00		
Item Description	**Quantity**	**Sr. Controls Engineer (Hrs./Ea.)**	**Jr. Controls Eng (Hrs./Ea.)**	**Total Hours**	**Budget**
Configure I/O Points	50	1	3	200	$16,250.00
* Sr. Controls Engineer Rate=$100/hr.					
**Jr. Controls Engineer Rate=$75/hr.					

Figure 4.2. Example budget headers for services.

For tasks that involve purchasing materials or products, you want to include fields for the quantity of items, unit of measurement, and cost per unit. When quantity is multiplied by cost per unit, the total material cost per the budget line item is calculated. A basic header for material budgets with a few examples is shown in Figure 4.3.

Total Cost Is a Field Calculated By Multiplying Quantity and Unit Cost

Item Description	Quantity	Units	Unit Cost	Total Cost
HART Enabled Analog Input Cards (16 points)	5	ea	$1,400.00	$7,000.00
Cable for Marshalling Cabinet to I/O Cabinet Runs	2000	ft	$3.00	$6,000.00

Figure 4.3. Example budget headers for materials.

As you create your budget for the migration project, you will want to establish a structure up front that simplifies the reporting process on the back end. There is not a single right way to organize your budget. Ultimately, the right approach is one organized in way that makes the most logical sense to you and is easy for you to track and adjust as needed. The budget is not only a platform to monitoring spending, but also a way of tracking task completion progress. Designed with this in mind, your budget spreadsheet can become a key tool that supports progress monitoring and reporting.

SCHEDULE

The schedule is a great crosscheck for your budget helping you to validate that budget assumptions are correct. There are a number of well-known and often utilized software tools (e.g., Microsoft Project™) for tracking project schedules. We will not focus on the specific scheduling tools but instead will review in simple terms the critical areas to consider in building a schedule along with some potential pitfalls.

While some companies use the scheduling software tools to generate their budgets, I recommend generating a separate budget and using the scheduling software to validate it. This validation process is a valuable and worthwhile exercise. If you use the schedule to generate your budget, any mistakes or incorrect assumptions will be repeated in both documents making the issues more difficult to identify.

Accurate project schedules include several elements related to each work task: work, start date, duration, predecessors, and resources. The work category defines the number of hours involved to complete a task. The start date identifies the calendar date that you want a task to begin. The duration defines the calendar time that a task will take in hours, days, or weeks. Predecessors identify any other tasks that must be started or completed before the successor task can either begin or end.

Resource fields can be used to identify the quantity of resources required to complete a task in a given duration as well as the discipline and experience level of the resources. The granularity of the resource category can be general or might include specific named resources depending on how you approach your project staffing. It should be noted that equipment and materials can also be defined as resources. In Figure 4.4 below, the pyramid reflects the increasingly granular categorization of staffing resources.

At the most general level, you may use generic categories for resources such as controls engineering. At the most granular level, you may choose to detail specific resources by name. Determining how to handle the resource

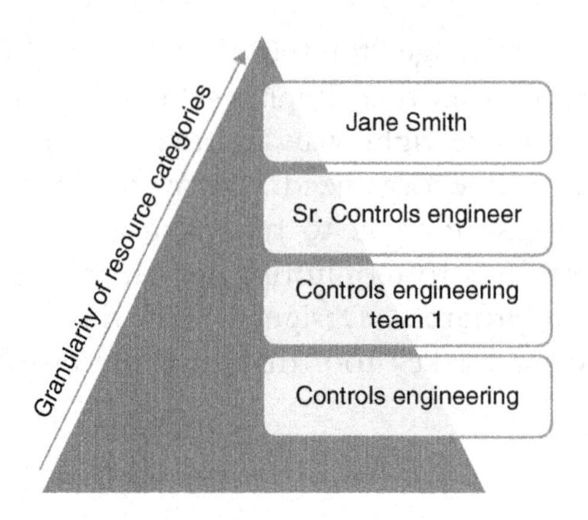

Figure 4.4. Schedule resource category options.

categories within a schedule is influenced by the many factors including the number of resources involved in your project, how the work will be executed, and the bill rate variability among resources within a given discipline. Highly granular resource categories can add to the complexity of managing your schedule and associated staffing plans.

A number of the basic categories we have outlined can also be used to calculate or generate additional fields. For example, you can set up resource rates so that as you apply the resources to task in combination with the work or duration factors, the cost can be calculated and included as an additional field. This can help you develop an accurate project staffing plan. Most schedule software programs have resource loading capabilities so that you can evaluate various staffing scenarios to find the optimum approach for your project.

I recommend keeping schedule tasks at similar resolution levels as your budget. There may be occasions when you want to build out some additional subtasks due to predecessor relationships or other factors such as a division of work task responsibilities among different resources. However, if you find the need to do this repeatedly, you probably do not have enough granular resolution in your budget. When developing your schedule, you must be very clear on the workflow and interdependence of work activities.

A common mistake that creates havoc for many projects is not identifying predecessor relationships. This is especially problematic when you have numerous resources involved in similar or related tasks. The example below shows the importance of building the predecessor and successor relationships into your schedule even if you think they are not significant.

Example

The project schedule for your control system migration includes a construction task to install and wire three new marshalling cabinets in the existing marshalling room. The task was not linked to any predecessors but did have a number of cascaded successor activities with very tight scheduling windows.

The construction team knows that a forklift will be needed for lifting and moving the marshalling cabinets to an area near the marshalling room. Based on the schedule and the criticality of getting the cabinets in place for the subsequent work task, the construction manager schedules the forklift. Two days before the work the construction manager checks the warehouse and realizes that the cabinets have not been delivered. Upon further investigation, it is determined that the marshalling cabinets were ordered later than planned and will not be ready for another week.

The delivery of the cabinets should have been linked as a predecessor task to the marshalling cabinet installation. If this had been done, then when the project manager updated the schedule to reflect the delayed ordering of the cabinets, the successor activity start dates would have been automatically shifted back. What was the impact of missing the predecessor relationship of these activities? The construction manager has to revise the work plans for his team as well as cancelling the scheduling of the forklift. In this case those things can be done without a cost penalty but that is not always the case.

As illustrated in the example above, the impact of small shifts in the start or completion of one activity can have a domino effect and lead to inefficient use of manpower and larger delays in more significant work tasks. Extended work durations are another area that frequently compromise schedules. The duration of activities can be impacted by many outside factors such as delayed equipment or material deliveries and operational issues that slow the work permitting process. Alternately, staff inefficiencies can also contribute to longer than expected work durations.

SUMMARY

A project's scope, schedule, and budget form the foundation for good project management. These documents should be organized at the beginning of the project to simplify project management activities such as progress monitoring, change order management, and reporting later in the project. Using each document to crosscheck the other documents for consistency is a good way to identify potential issues in your project. The scope of work for a control system migration project has a number of functional disciplines that must be addressed. While the scope in some of these areas varies greatly based on your project specifics, the major tasks in each discipline area should be evaluated for applicability as part of assessing whether you have a complete project scope.

When establishing a budget, you want to strike a balance between manageability and the right level of detail. Due to significant labor services required for migration projects, your budget should accommodate variable rates to improve budget accuracy and simplify management of changes when the project is in progress.

Your schedule tasks and work activities should be at a consistent granular resolution with your budget. However, there may be some additional detailed subtasks needed to account for predecessor relationships and segmented calendar work activities. When establishing your schedule, be especially mindful of capturing predecessor and successor relationships between tasks.

Three Key Takeaways

- A project's scope, schedule, and budget have a fixed relationship and should be closely aligned with consistent task-level resolution.
- The scope of work document should address all areas of the project in sufficient detail to enable the project manager to build an effective work breakdown structure that will be used as the foundation of the budget and schedule.
- Budgets for projects with heavy labor services are most accurate and easy to update with changes when they include a way of tracking and accommodating variable bill rates.

5

Project Staffing

One of the most unique things about a control system migration is the diverse staffing mix required to achieve a successful project. Migration project teams are often comprised of blended resources from a variety of disciplines, potentially several different vendors, and multiple departments of the end user company. A balanced team with a clear understanding of individual responsibilities and the ability to effectively communicate is critical to a successful migration project. Some key characteristics to look for in your project team members are discipline area expertise, experience level, communication skills, and attitude. Often it is the latter two capabilities that prove most valuable in building an effective team.

Outside of the assigned project team members, it is also important to identify key individuals in other parts of the end user organization who can help support the project and provide valuable input. The effectiveness with which the project team interfaces with key departments (e.g., operations) will influence the amount of input and the level of cooperation the team receives. This can improve project efficiency and result in a better overall control system design and implementation.

The project execution strategy largely defines the project staffing process. For example, if an EPC company is executing the project, then the end user company has a small role if any, in individual staffing decisions. When staffing a control system migration project, it is worth considering a potential resources experience with the existing control system and knowledge of the new control system. A thorough understanding of how the existing control system is configured and operates is essential in establishing the functionality needs of the system and making the transition to the new control system easier. Much of the in-depth knowledge of legacy control systems has been lost through retirement and general attrition. This is particularly a problem when an EPC company, vendor, or system integrator is staffing the project because they are often at the disadvantage of not only a lack of familiarity with the existing control system, but also a lack of understanding of the production process. This combined knowledge deficiency can make it very challenging to accurately grasp the detailed functionality of the existing control system. Identifying resources within the end user company knowledgeable on the configuration and operation of the existing control system who can either become members of the formal project team or consult with the project team can offset these deficiencies and be a key migration project success factor.

Alternately, a lack of knowledge of the new control system by either the end user company or third-part service providers can also create challenges. The learning curve on the new system can extend the project schedule. Also, the discovery or improved understanding of new features and functionality as the project progresses can result in rework or cause inconsistencies in system configuration. To minimize issues in this scenario, it is recommended that someone on the project team have expertise on the new control system. This might require the end user to hire the vendor or a system integrator with knowledge of the new control system to supply consulting or configuration support services if they are not going to be the primary EPC service provider.

In this chapter, we discuss how to define your staffing needs and develop clear roles and responsibilities on the project team. We also outline how to extend the team concept to those outside of the core project team and why this is important for project success. Establishing frequent and inclusive communications is another key element that we examine. Finally, we will review ways to emphasize teamwork and build a stronger control system migration project team.

DEFINING PROJECT RESOURCE REQUIREMENTS

The scope of work document identifies discipline areas that are essential to your migration project and serves as the foundation for creating your project staffing plan. When used in combination with your budget and schedule, the type and number of resources needed for your project will begin to take shape. Once project staffing needs are identified, a team can be assembled and individual team member roles and responsibilities can be defined. Each team resource needs to understand not only their specific assignment, but also be familiar with the responsibilities of other team members. Understanding their own responsibilities is obviously necessary to complete their work, improves efficiency, and promotes individual ownership within the project.

Establishing a familiarity with the responsibilities of other team members encourages proactive information exchange, speeds resolution of issues when they arise, and helps with transitioning work activities across responsibility areas.

The three core tools that a project manager uses to help define project staffing needs and outline work responsibilities are as follows:

1. Organizational Chart
2. Roles and Responsibilities Matrix
3. Schedule

In the sections below, we will explain more about each of these tools including their purposes and how they can be successfully used for your migration project.

Project Organizational Chart

A project organizational chart enhances the understanding of how all members fit into the overall team and identifies a hierarchy of project responsibilities. Just as important, the organizational chart communicates the project team structure to those external departments and individuals with which the team will be interacting as part of the project. Many control system migration projects do not develop organizational charts as part of their project management documentation. There are times when it can be omitted such as on smaller projects or those that are primarily executed by a single entity as opposed to blended teams. However, these are exceptions, and the benefits of developing an organizational chart are worth the time invested. A sample organizational chart is provided in Figure 5.1 below.

The degree of appropriate detail in a project organizational chart is dependent on a number of factors. For example, if the project is minimally staffed you may want to include the entire project team, whereas on more complex projects you may only show a few reporting levels. I typically suggest defining resources at least three levels deep. I also recommend using individual names in addition to titles in the organizational chart for the sake of clarity. The project organizational chart can serve as the master document that links job titles with individuals. Ultimately you should make any adjustments to the project organizational chart that you feel make it the most valuable and effective documentation and communication tool for your project.

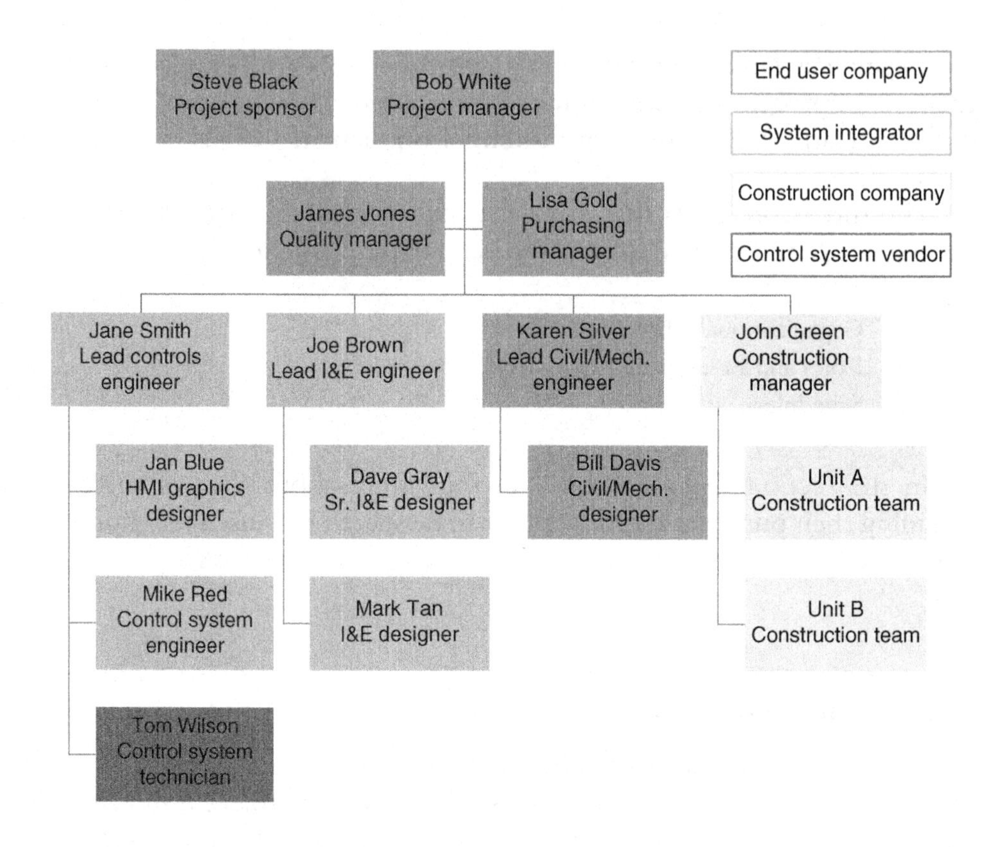

Figure 5.1. Sample project organizational chart.

Roles and Responsibilities Matrix

Another frequently used tool to help define project staffing is the roles and responsibilities matrix. The roles and responsibilities matrix identifies key project tasks or responsibilities on one axis, job functions, or titles on the other axis and defines roles at the intersection points. In this case, role refers to the function that an individual serves for a given task as opposed to a description of their discipline area and title. The detail you use to define both work tasks and job functions is at your discretion. I suggest using major work task categories as opposed to detail subtasks. For example, I would use graphics configuration as a work task rather than Unit 1 graphics configuration, Unit 2 graphics configuration, etc., unless different individuals are responsible for the graphics in each unique area and you are trying to make that designation known.

I also recommend using broad discipline titles and not individual names as a rule of thumb in the roles and responsibilities matrix. For instance, rather than Joe Smith I might use Lead Controls Engineer as the designation. If resource changes occur during a project, this will minimize the require documentation revisions necessary. The organizational chart can serve as the master document linking the job titles with specific named individuals.

Once the work tasks and job functions within your roles and responsibilities matrix are created, you need to identify the role each party has for a given task. One of the most frequently used definition processes is the RACI model, which I have outlined in Table 5.1 below.

Table 5.1. RACI method definitions

Role	Responsibility Definition
Responsible (R)	Performs the work required to accomplish a task.
Approver (A)	Assigns responsibility for a task and signs off that a task is completed and acceptable.
Consulted (C)	Provides subject matter expertise, input and guidance regarding a task.
Informed (I)	Aware of the status of a task but not involved in task execution.

The RACI method is straightforward and easy to understand making it a default for many projects. There are numerous variations of this model as well

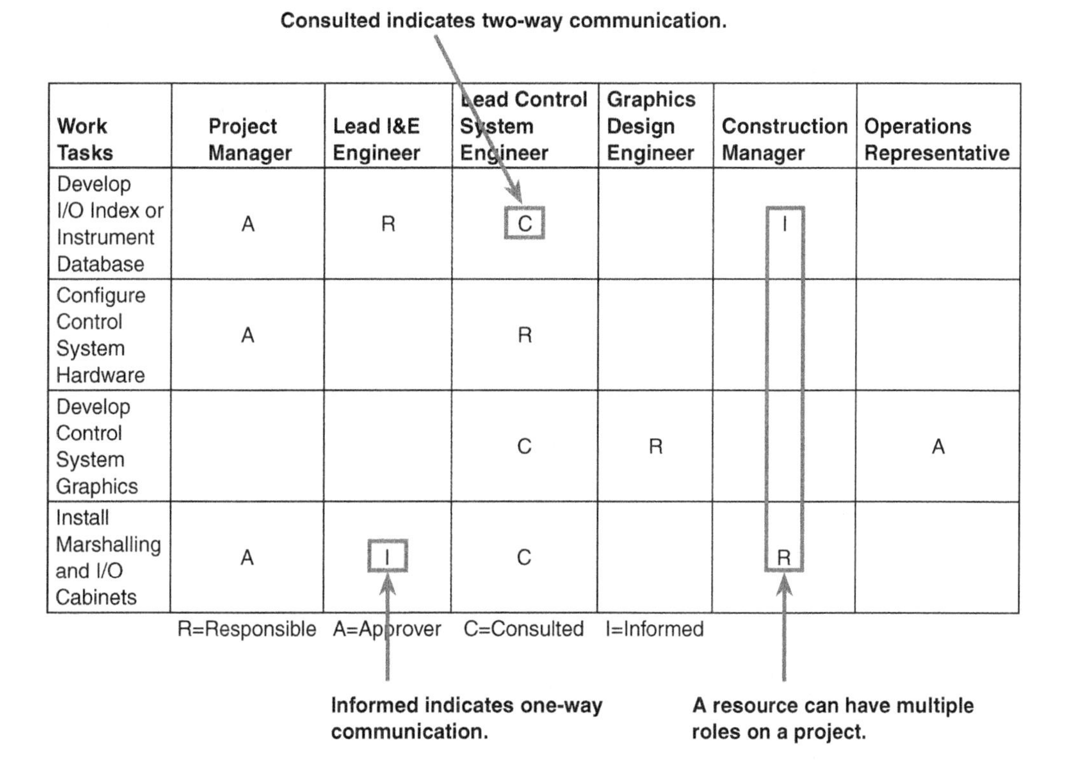

Work Tasks	Project Manager	Lead I&E Engineer	Lead Control System Engineer	Graphics Design Engineer	Construction Manager	Operations Representative
Develop I/O Index or Instrument Database	A	R	C		I	
Configure Control System Hardware	A		R			
Develop Control System Graphics			C	R		A
Install Marshalling and I/O Cabinets	A	I	C		R	

Figure 5.2. Example partial roles and responsibilities matrix.

as other responsibility definition models, which can be applied to the roles and responsibilities matrix. Regardless of the definition method selected, the roles and corresponding responsibilities definitions need to be clearly understood by those involved in the project.

Figure 5.2 above is an example section of a partial roles and responsibilities matrix to illustrate the concept.

Note that in the case of this example, the roles are only defined to the level of discipline leads. For example, a control system designer might actually do the design work associated with the I/O and marshalling termination schedules, but the lead control system engineer is still listed as the responsible party. You have the option to drill down further and might include the control system designer level role as the responsible party. However, it is a common practice to list the leads for each discipline category as the responsible party and then let them work out the details of specific project tasks among their discipline teams, while they maintain ultimate responsibility. This approach prevents the roles and responsibilities matrix from having so many resources listed that it becomes cumbersome.

Project Schedule Resourcing

The final staffing tool commonly used by project managers is the project schedule, which can also double as a detailed staffing plan. The project schedule's work breakdown structure is granular enough to enable resource assignments at the subtask level using the resource field. When naming resources within your project schedule, as mentioned in an earlier chapter you can choose resource names for work discipline groups, work teams, job titles, or specific named individuals. Whichever resource naming strategy you use, at the primary task level the associated resource assignments should be consistent with those listed in the roles and responsibilities matrix.

The flexibility of most scheduling tools gives you the ability to easily review and adjust assignments and track resource loading by individuals or teams. The work hours, work duration, and resources should be tracked for consistency against the budget and schedule throughout project execution as part of normal project monitoring activities. This can help you quickly identify resource loading issues that arise during the project.

EXTENDING THE PROJECT TEAM

There is a core project team responsible for completing the work scope tasks associated with the control system migration project. However, there are also numerous other departments and individuals integral to your project's success. If they are not included in the project design, decision-making process, and resolution of issues, the project will suffer.

The individuals within a company who have interest in a control system migration project normally include departments such as Operations and I&E Maintenance at both the management and technician levels. Other departments such as IT might be engaged as well. The project manager should work to keep those who were a part of the control system selection team involved as the migration project progresses. The benefits of their involvement include better designs, identification and elimination of potential issues up front, and greater cooperation during project execution and startup. The anecdote below emphasizes the importance of taking an inclusive approach to control system migration projects to avoid misperceptions and misunderstandings.

Anecdote

We were awarded a control system migration project on which our company partnered with a construction team that we had worked with successfully in the past. We chose to partner with this particular construction company because they were reasonably priced compared to some other companies we explored. Even more important, we were confident in their work quality and ability to work with our team. However, they had not previously worked at this particular customer site.

When the construction phase was starting to ramp up, we quickly recognized that we would have some challenges. Our client's I&E department had forged a great relationship with a local I&E construction contractor that the site regularly used. We had considered this I&E contractor as a potential partner for the project, but their price was substantially higher and we had no work history with them, which caused us to be hesitant to partner on a firm price bid. The resentment from our client company's I&E team was obvious from the time our construction team arrived at site.

The construction manager and I decided that the only way we could effectively build cooperation was to find ways to engage the client company's I&E team as part of the project. By doing this, we hoped they would have a vested interest in seeing the project succeed. Working with the client project manager, we requested and were assigned two of our client's I&E technicians to support the project. They were responsible for reviewing our I&E construction work, providing support, and helping manage the cutover for the client.

Once they began working with our team, our construction I&E team started asking them questions and seeking their advice. The client's I&E technicians quickly developed relationships with our team and began praising our construction team to their counterparts when they were back in the I&E shop. The resentment almost immediately died down and the project proceeded without incident. The company I&E team became great allies and were extremely helpful in leading the project cutover and supporting our construction efforts.

New working relationships, biases, personality conflicts, and other people-related obstacles are to be expected on every project. Good project management includes recognizing these challenges and implementing proactive solutions to avoid or minimize the impact. Creating an inclusive environment, which welcomes and values the perspective of not only those on the project team, but also others who have a vested interest in the project outcome is an important part of addressing these challenges.

We have discussed why it is important to have these various external departments involved in the project. However, in most situations, these resources have other full-time responsibilities and may have little or no time to contribute to the project. This can make getting them involved difficult. Holding regular project status update meetings with an extended team of key contributors is one key way to get and maintain involvement. I also recommended establishing a project advisory team as a good way of involving others in your migration project. Before you do this, make sure you are going to utilize the input of this team and are not only doing this for show. If you establish the team and call regular meetings that have no value, then you have wasted everyone's time.

Start by working with the team to create a clear charter. For instance, the advisory team's charter might be to review the project progress against project scope, schedule, and budget documentation at defined regular intervals throughout the project. Alternately, you might have the team evaluate project-related issues and recommend solutions, particularly those that are related to company policies, personnel, or operations. The advisory team is also a good group to challenge project decisions and strategies because of the diverse perspectives from throughout the organization that are represented.

By establishing an advisory team or other such group, the various departments that are stakeholders in the project stay familiar with the project and know when certain activities are planned. This enables involved departments to be more responsive to project needs. The core project team should also engage the advisory team members to assist in resolving issues and to champion project cooperation within their own departments.

There are of course logistical challenges related to getting a diverse group of individuals from different parts of the organization together. Flexibility is usually required on behalf of all team members when it comes to involving other parts of the organization that might not be dedicated to the project full-time. The following anecdote illustrates the logistical complexities often faced when involving others in control system migration projects.

Anecdote

For a control room consolidation project I was managing, the Operations Manager committed to provide an operator from each shift to consult with us on a new control room layout, new control room furniture, and various details related to operational graphics. The team meetings needed to include myself, the controls engineering staff configuring the system, and four operators representing each of the shifts.

We were asked to minimize overtime and additional operator staffing requirement in our planning of meetings and other activities with the team of operators. In order to accommodate these requirements, we scheduled meetings with the entire group on the days when the shifts were changing and we could minimize the impact to the operators. We took advantage of days when off rotation shifts were already onsite for scheduled training. We planned meetings in a roughly two-hour window around shift change on select days. The controls engineering staff and I would stay late for these meetings. This schedule required flexibility and sacrifices on everyone's part. The operations representatives actively participated in the meetings and had a significant voice in setting the direction for the updated control room.

There are times when you will seek involvement from individuals who are repeatedly too busy to attend a meeting or otherwise participate. This can be frustrating and lead to project delays. The first step in resolving this issue is to speak with the individual to see if there are ways to adjust the schedule to better accommodate their availability. For example, scheduling a meeting at 7:00 a.m. and asking the operations shift supervisor to attend at that time might be difficult since that may be too soon after shift change.

In other circumstances, you may have individuals who are just never going to prioritize the project. In these situations, you should consider whether you really need that specific individual involved or if someone else might be a better choice to provide similar perspective and input. If it needs to be that individual, one way to encourage participation is to circulate the meeting agenda, meeting minutes, and meeting attendance information to management after each meeting. Most people take notice of the distribution list of this information and if their management is being copied on regular distributions that document their lack of participation, it usually motivates them to find time in their schedule.

ESTABLISHING TEAM COMMUNICATION

Good communications are essential to project success and facilitating effective communications among team members is a key project management responsibility. On the surface, this may seem like an easy task, but it can actually be a significant challenge. Intuitively, we all understand the importance of communicating with others. On migration projects, there is information and knowledge that must be shared and transferred among different individuals and discipline areas. This can include technical information, status updates on work activities, timing of planned activities, etc.

Unfortunately, communications can sometimes be a weakness for many engineers and technical oriented people. When communications on a migration project are not good, many of the details in areas where one individual's responsibilities end and another's begin can become misaligned as reflected in the example below.

Example

Your project lead control system engineer is a younger engineer with technical savvy far beyond his years. You have the utmost respect for and trust in his technical abilities. The lead designer on the project is extremely detail-oriented and very experienced albeit with a bit of a quirky personality. As your project progresses you identify some conflicts between some addressing assignments and control loop designs versus their actual implementation in the control system.

Upon investigating your lead designer says that he did not verify the addressing assignments with the control system engineer because he was never available. You talk with the control system engineer and he says that the lead designer never asked him to verify the addressing assignments. You pull both parties into a room to try to understand where the communication breakdown occurred. You determine that the young engineer has headphones on listening to music while he works. The lead designer took this as time that the engineer was unavailable and considered the young engineer unapproachable. As a result of this misunderstanding, you begin calling regular staffing meetings to facilitate communications between all parties.

The above example may seem like an exaggeration of a communication issue, but it is somewhat anecdotal from one of my projects with some minor changes. People are all different as are their respective approaches and perspectives. The types of challenges outlined in the example above will happen and project managers must be actively engaged in the people management aspect of their job responsibilities to help avoid issues such as these and to quickly respond when they do occur.

Build time for regular team meetings into your budget and schedule. The first meeting that needs to be held is a team project kickoff meeting. This meeting is essential for reviewing the project scope, schedule, and budget with the team and establishing project protocols and expectations. Meeting time can often be seen as an inefficient use of valuable time by both team resources and company management. That is only the case if the meeting is not managed properly and from my experience this is too frequently the situation.

Project meetings promote the exchange of information, improved awareness of project status, and a common understanding of priorities. Requiring each discipline team lead to review the status of their activities, discuss planned priorities for the upcoming week, and identify any issues or areas of concern encourages them to manage their work scope and better track their progress against scope, schedule, and budget. Regular team meetings with clear agendas, designed to require team member participation improve team communications. Figure 5.3 below is an example of a control system migration team meeting agenda, which uses broad agenda subjects to illustrate what types of topics should be covered. The other key concept reflected in the example is that each discipline lead has an active participation role in the meeting.

Control system migration meeting agenda
week of March 11, 2013

Resource	Duration	Subject
Project manager	10 minutes	Safety review
Project manager	15 minutes	Progress review of overall scope, schedule and budget
Lead I&E engineer	10 minutes	Update on status of I&E scope, planned work for week, known issues
Lead controls engineer	10 minutes	Update on status of controls scope, planned work for week, known issues
Construction manager	10 minutes	Update on status of construction scope, planned work for week, known issues
Project manager	5 minutes	Identify any outstanding issues and areas needing PM assistance

Figure 5.3. Sample control system migration project team meeting agenda.

Another way to strengthen team communication is by designating regular reporting. Project managers are required to provide project reports at some regular interval such as weekly. Each team lead should also be required to provide a discipline report that the project manager can roll up into their overall report. The written report that the discipline lead provides does not have to be formal or lengthy but must be timely and accurate.

The project manager should clearly define what information is to be reported. Beyond that I encourage project managers to keep it simple and informal. For example, emails can often suffice rather than some complex reporting form. The benefit of implementing a discipline reporting process is that it forces each lead to step back from the details of their work and take a bigger picture view of their overall work scope at regular intervals. Many times this is where issues are first identified.

Outside of meetings and reporting, the project manager's level of engagement is also a vital part of driving project communications. Project management is about more than just filling out forms and reports, maintaining a schedule or tracking to a budget. To be effective, project managers must spend time visiting key project resources, asking questions and completing first hand assessments of progress. This will help the project manager build trust with team members and promote better communications.

BUILDING AN EFFECTIVE TEAM

Every project manager wants the most talented team members available assigned to their project. However, experienced project managers understand that a collection of the best technical resources does not automatically translate into an effective team. While it is important to have individual team members who know their areas of expertise and can effectively perform their individual job functions, it is also vital that the team be able to function as a cohesive unit. Technical resources tend to be data oriented and analytical but sometimes struggle with communications. Strong teams share information and knowledge, are responsive to one another, and are willing to extend the boundaries of individual responsibilities when necessary to cover the gray areas and transition zones of a project.

The blended teams resourcing many control system migration projects make team building efforts a necessity. It is unrealistic to expect a team to hit the ground running and work efficiently together when they have never met until the project. Building relationships and trust between team members at the beginning of a project increases team efficiency and cooperation, strengthening the chances for overall project success. The anecdote below recalls a case where team building exercises were a valuable tool in creating a more cooperative team environment.

Anecdote

Our company had recently completed a large control system migration project and while successful, we had encountered communication breakdowns and high levels of frustration from several team members. When we were awarded another, even larger migration project, it became obvious to me that we would likely repeat some of the same mistakes if we did not approach the project differently.

The team for the new project included several members from the first project as well as some employees who were relatively new to our company. I took a small portion of the project budget and rented a space at the local college to hold an internal project kickoff and team building meeting. I purchased a book that included numerous team building exercises and selected a couple in which I thought the team would be willing to participate. Some of the exercises required the team members to answer questions prior to the meeting. They were intrigued but a bit weary of what we would be doing in the meeting.

On the day of the meeting, we executed normal kickoff activities such as reviewing the scope, but also included several of these team building activities. At first they were hesitant participants but toward the end were fully engaged and enjoying these exercises. The team became more relaxed with one another and began to recognize that some of the past issues on the last project may have been misunderstandings. It motivated teamwork and pushed team members to give each other the benefit of the doubt. As a result, the communications and teamwork on the project were substantially better than on the previous project.

You might be rolling your eyes at the thought of cheesy team building exercises. I completely understand the hesitancy and apprehension, but the human aspects of a project are as important as the technical aspects to the overall probability of success. An individual may be able to adequately perform their scope in isolation but it won't necessarily be in the best interest of the project. Technical knowledge and expertise are only valuable when they can be successfully applied within the context of a team to support the overall project goals. The example below demonstrates how a lack of communications and teamwork can result in project issues.

Example

Your project requires several field instruments be replaced. The old conduit is having some corrosion issues so new conduit will be required from the field junction box location. The senior designer on the project designates the conduit routing as part of the design and turns over the design package to field construction. When the field construction team begins the work, they run the conduit as designed. As they are preparing to finalize terminations, the field construction manager comes by to review the installation and recognizes that the conduit is routed too close to another piece of equipment and will impede maintenance access. The conduit must be re-routed costing additional money and time.

This is an example where the designer completed his work task as designated. However, communication between the designer and the construction manager in the design stage or during a constructability review would have benefitted the project. A project manager cannot force all communications on a project. It is essential that team members be able to work cooperatively together initiating communications and discussions as part of daily work activities.

What is it that we are trying to accomplish by focusing on team building? We want to eliminate barriers to knowledge sharing, communication, cooperation, and mutual decision-making. These are vital aspects of team performance. For example, if you have a senior designer who wants to be the expert on the project so he isn't willing to share insight or information with another designer, the second designer's efficiency will decrease and the design will also likely suffer.

Outside of team building exercises, the suggestions made in the communications section of this chapter such as regular meetings will also support improved teamwork. Team incentives are another way to encourage strong working relationships on the team. These can take many forms but in the end if the team understands that they are judged as a group, they are more likely to emphasize the team approach.

SUMMARY

Control system migration projects usually require diverse staffs from a variety of sources. Establishing a strong team requires selecting the right individuals, making sure they are clear on their roles and responsibilities, facilitating communications among team members, and encouraging teamwork. There are other parties who are also important for project success and finding ways to engage and include them in the project has numerous benefits.

The staffing process begins by defining the resource needs based on the scope of work document. Project managers can then use a combination of tools to assign responsibilities and communicate those to both the team members and others who will interact with the team during project execution. Regular team communications should be emphasized and team building efforts at the beginning of a project can often result in improved teamwork. Knowledgeable and cooperative resources functioning as a team are vital to executing a successful control system migration project.

Three Key Takeaways

- A balanced team with a clear understanding of individual responsibilities and the ability to effectively communicate is critical to a successful migration project.
- Three core project management staffing tools are a project organizational chart, a roles and responsibilities matrix, and the resource field of the project schedule.
- Establishing an advisory team or otherwise involving stakeholders from key departments that will be impacted by the project can contribute to improvements in project design, streamlined decision-making and approval processes, and enhanced overall responsiveness.

6

Training

Training should be a high priority for the project manager of a migration project because knowledge of the new control system is critical for cutover efficiency, successful project turnover, and safe operations. There are a number of complex aspects to training on a new control system that should be addressed as part of a comprehensive training plan. The long-term successful adoption of a new control system is fundamentally rooted in good upfront training as a part of the migration project.

Training is typically an integrated effort requiring involvement by multiple individuals from the project team, the control system vendor, the company training department, and individual departments such as I&E maintenance and operations. Coordinating input and reaching consensus on how to provide the most effective training can be challenging. The project manager needs to take responsibility for ensuring that a training plan is developed, which both supports the immediate project needs and facilitates long-term operational success.

Funding training efforts associated with a new control system can also be complicated. Depending on the number of employees to be trained, the types of courses used and the location of the courses, training costs can be substantial. In some companies training is considered part of the project and therefore included in the project budget. However, other companies fund training with the individual departmental budgets or a separate training department budget. The project manager needs to be clear on what training is needed and how it will be funded from the inception of the project. Even when the training budget is handled independent of the project, the project manager has a vested interest in making sure training efforts are successful and should be a central contributor to developing the training plan.

Developing a complete training plan requires not only defining content needs, but also considering multiple aspects of training logistics. Effective training plans document:

1. Who Should Be Trained?
2. What Content Should Be Covered?
3. When Training Should Occur?
4. Where Training Should Take Place?
5. How Training Should Be Executed?

In the following paragraphs, we examine each of these in more detail.

Training requirements vary widely among different industries, but fundamentally anyone who has regular involvement with the control system needs to clearly understand how to perform their job responsibilities with the new system. Training for migration projects includes a variety of personnel from the operations, maintenance, and engineering departments. As a result of the different needs of these departments, training associated with control system migration projects requires multiple courses covering diverse content and often includes several layers of knowledge depth addressing different levels of expertise required.

The timing of training is a key factor in determining the degree of success. The goal is to maximize retention rate. If you train too early in the project without resources having a system available, the retention rate will be low and additional refresher training may be required shortly before transitioning to the new system. Alternately, holding training too near to the transition can create a rushed and higher pressure learning environment, causing resources to feel uncomfortable with their preparedness.

There is no universal answer for when to train everyone. Different department needs and specific roles of individual employees are important considerations that help determine the most appropriate timing for training. For instance, engineering training is typically completed early in a project, while operator training is typically done late in a project. Ultimately, training timing is somewhat dictated by a combination of training course or instructor schedules, trainee availability, and other logistical factors.

Location is another key aspect of the training program and includes the option of training at the site, at a vendor hosted location, at a local offsite venue or online. The training environment will impact the effectiveness of the training and should be adequately considered as a valued component of the training experience. If you are training only a few individuals, as is the case with engineering training, then attending training at the vendor hosted location is a practical option. This option is typically not practical for large groups of trainees such as operators because of the substantial travel and living costs. The exception to this is when the vendor hosted location is local.

While management may prefer onsite training when possible, a common complaint about onsite training is that there are often distractions and interruptions that can lessen the learning experience of attendees. The advantage of onsite training is that it reduces travel and living expenses as well as minimizing productivity lost due to travel time. This makes it a practical solution for a large number of trainees. The example below highlights the challenges of holding onsite training.

Example

Control room operator training is being held in a training room on the plant site. The migration project manager has been assured that the operators will not be interrupted during the training classes. During the morning break, one of the shift supervisors attending the training starts a conversation in the hallway with the operations manager. One of the production lines is down and the operations manager asks the experienced shift supervisor to stop by the control room and see if he can help determine why the line shutdown. The shift supervisor does not return to the training class until well into the afternoon session.

The example above shows that even with the best intentions to limit distractions and interruptions, they can easily occur when training is onsite. When training needs to be local, using offsite facilities is the preferred approach to minimize distractions and interruptions. There are often rental fees associated with offsite space, but they are usually reasonable.

Another option used more frequently these days is online training. There are a variety of approaches and delivery methods for online training. Online training courses can be effective in certain situations depending on how they are executed. Look for vendor-provided, online courses that are instructor led. The courses should be interactive with opportunities for questions and answers. These courses can have lower costs and also minimize travel and living expenses. Disadvantages of online training are that it can result in a less comprehensive learning experience and trainees can be susceptible to distractions and interruptions.

Table 6.1 below summarizes the different training location options and identifies common advantages and disadvantages to each.

While most project managers are not experts on organizational learning, it is important to have a basic understanding of learning and retention models. This will help with the process of determining which course designs and delivery approaches should be used as well as how to time the training for

Table 6.1. Summary of training location options

Training Location Options	Advantages	Disadvantages
Vendor hosted location	• Established and working equipment for hands on work • No equipment shipping expenses • Vendor responsible for logistical items • Minimal distractions or interruptions	• Travel and living expenses • Lost productivity for travel time
Local, onsite	• Travel and living expenses for instructor only • Trainee convenience • No lost productivity for travel time	• Equipment shipping expenses • Setup and checkout of equipment • Company responsible for logistical items • Common to experience distractions and interruptions
Local, offsite	• Travel and living expenses for instructor only • Trainee convenience • No lost productivity for travel time • Minimal distractions or interruptions	• Equipment shipping expenses • Setup and checkout of equipment • Company responsible for logistical items • Fees for facility usage
Online	• No travel and living expenses • Trainee convenience • No lost productivity for travel time • Course costs often lower • No equipment shipping expenses • No logistical responsibilities	• Common to experience distractions and interruptions • May not be as comprehensive of learning experience

maximum knowledge retention. A common model that was first introduced in the 1960s and is still frequently cited today is known as the learning pyramid. A depiction of this model is shown in Figure 6.1 below.

The pyramid indicates that retention is improved through approaches that engage students and encourage active participation. Courses designed by vendors typically take these learning and retention concepts into consideration and are optimized for productive learning. However, because of this it is often

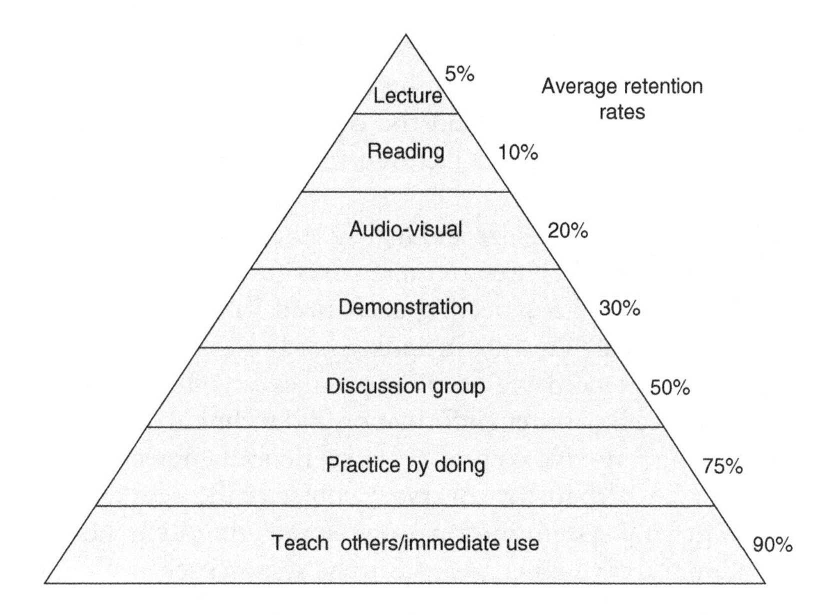

Figure 6.1. Basic learning pyramid.

difficult for these courses to be customized. Any customization requirements for training associated with your project need to be identified up front and weighed against their impact on the course cost and effectiveness.

In the remainder of this chapter, we will identify the typical engineering, maintenance, and operations department training needs on control system migration projects. We also examine others who might require or benefit from training on the new control system. Finally, we will review commonly encountered training challenges and outline recommended ways to address and enhance training for the various departments.

ENGINEERING

Comprehensive training on the new control system within the engineering group is typically isolated to the controls engineers or others responsible for configuration of the control system. This training is the most extensive and in-depth as the individuals attending this training are the long-term caretakers for the control system. Because these engineers are often heavily involved in the project and require a thorough understanding of the control system to make design decisions, this training should be done near the beginning of the project or even prior to starting the engineering phase of the migration.

Engineering training normally consists of attending one or more vendor training classes at a hosted training location. However, many online courses are also now available and can sometimes be used as a substitute for vendor location training. The costs of engineering courses vary based on the particular vendor, the length of training, and the course content. Travel and living expenses should be anticipated and factored into the budget associated with many of these courses.

While different vendors organize and manage their training programs differently, there are fundamental content requirements for good engineering training. At the most basic level engineering users need to understand the system terminology and system architecture including hardware components and software applications. Experienced engineering resources can often acquire this basic knowledge through reading system information and technical documentation as opposed to attending a specific course. Less experienced engineering resources might benefit from an introductory overview course of the new system.

Beyond this general system information, engineering users need to learn how to setup administrative functions within the system such as user accounts, alarms, system security, etc. They also need to learn how to configure the hardware and software within the system. Most vendors offer hands on introductory courses that will provide enough detail to cover these aspects of system engineering and configuration. It is important that these courses give engineers the opportunity to actively configure the system as oppose to observing, to engage them in the learning experience.

Additional courses that provide more in-depth drill down on specific topics might also be needed depending on the experience level of the controls engineering staff and the complexity of the control system. Advanced configuration courses for graphics as well as function block and logic programming

Table 6.2. Engineering training recommendations

Training Questions	Common Recommendations
Who should be trained?	Process control engineers
What content should be covered?	• System terminology • System architecture • Communications • System administration configuration • Hardware component configuration • Software configuration (points, graphics, logic)
When should training occur?	Prior to or at the beginning of the migration project
Where should training take place?	Vendor hosted location
How should training be executed?	Standard vendor courses with hands on opportunities

are some common examples. If any specialized functions such as batch or advanced process control are being implemented, additional training may be required as it is common for vendors to offer individual courses that drill down into these topics.

At the completion of training for the control system engineering team, these trainees should be the experts within the company capable of configuring points, graphics and logic in the system, optimizing tuning and other parameter settings, troubleshooting complex issues, and answering system questions for the operations and maintenance groups.

Table 6.2 above provides a summary of recommendations for engineering training.

MAINTENANCE

Training for I&E maintenance staff is often overlooked as a critical requirement for control system migration project success. Unfortunately, many organizations seem convinced that the instrument and electrical aspects of one control system is similar enough to the next that the I&E staff can pick it up and adapt without needing formal training. Training I&E staff on the new control system will avoid inefficiencies and potentially costly mistakes.

Training all I&E maintenance technicians is not always feasible. An alternate approach that can be effective is to train several key technicians on the new system as the resident experts. They are then responsible for disseminating the training knowledge to the other technicians. This is often referred to as the "Train the Trainer" approach. The FAT and SAT also present excellent opportunities to involve I&E technicians getting them first-hand experience with the new control system.

Training of maintenance staff should cover system terminology and system architecture including hardware components and software applications at a basic level. The focus of maintenance training should be on how key components of the system function and how to use the system diagnostics along with other available information to troubleshoot issues. Control system vendors typically offer standard courses focused on I&E maintenance training, which are a combination of workshops and lectures. Customization of these courses is not needed and they are generally offered at vendor hosted training locations. If you have a large staff of I&E maintenance technicians to train some vendors will hold onsite courses. Keep in mind that these maintenance workshops require functioning equipment, which can sometimes create logistical challenges and require install and setup time for courses to be held locally.

At the end of maintenance training, I&E technicians should be:

- Able to access information in the system such as basic point configuration information or controller faceplates from graphics.
- Knowledgeable on the system architecture, wiring, power, and grounding philosophies.
- Comfortable troubleshooting basic issues.

Table 6.3 below provides a summary of recommendations for maintenance training.

Table 6.3. Maintenance training recommendations

Training Questions	Common Recommendations
Who should be trained?	I&E maintenance technicians I&E maintenance supervisors and managers (optional)
What content should be covered?	• System terminology • System architecture • Hardware components • Software applications • Communications • Wiring, power and grounding philosophies • Basic point, loop, and logic configuration from graphics displays • System diagnostics • Troubleshooting tips
When should training occur?	Early in the project, no later than mid-project if they are providing cutover support
Where should training take place?	Vendor hosted location
How should training be executed?	Standard vendor courses with hands on opportunities

OPERATIONS

Training for the operations department is typically the most complex. There are a large number of operators that must be trained with logistical challenges to overcome. Operations teams are on a scheduled shift rotation, so finding a coverage schedule that supports both operating the facility and training all operations shifts is not always straightforward. There are also time constraints to the availability of operators, which makes course duration a key consideration.

Several organizational levels within the operations department may require training. The control room operators responsible for controlling process operations will need to learn how to access and navigate displays, monitor process performance, and view real-time and historical trend information. They also must learn how to manipulate a variety of control parameters and respond to process alarms. Ultimately, training for these operators must provide them with the knowledge to use the tools within the new control system to access information, make decisions, and adjust control parameters as needed for process operations.

A second layer within the operations department that may want training is management at the shift supervisor level or higher. Field operators may be a third layer within the operations department requiring training. The training for both management and field operators frequently covers how to access and navigate displays, call up individual points and alarm information, and view trends.

There are a variety of ways to execute operations department training. The control room operator's course is typically a vendor supplied course. Many companies choose to customize this course so that there operators are learning navigation and other functionality specific to their own operations. Application-specific training will increase the training costs but may be a much more effective training approach to increase the operator comfort level, improve retention, and make the transition easier. For those that use the general course, at some point you will also want to provide supplementary documentation or additional training that outlines your application-specific control system configuration including graphics and various operational functions.

The training for management and field operations can also be handled in a number of ways. These resources can attend the control room operator course and have full access to the same information as a control room technician. However, this is often costly and because they are not using the information as regularly, retention may not be strong. An alternate method that is frequently used with reasonable results is customizing documentation or training that provides an overview of the graphics hierarchy structure with associated navigation details. This training can be provided by the project team as opposed to the control system vendor reducing the cost. The goal is to provide instructions on how to do basic operational tasks such as calling up faceplates, accessing alarm summary screens, or reviewing trend data specific to your configured system.

The timing of operational training is a frequent subject of debate in control system migration projects. Operational training, if done too early, will be long forgotten by the time project cutover occurs and operators will need to be retrained. The general recommendation is to time the operator training within a few weeks of cutover if possible. Taking advantage of any opportunities to expose operators to the new control system can be beneficial. For instance, having several operators participate in the FAT can be a great way to enhance their knowledge of and comfort level with the new system while also making a valuable contribution to the project.

One alternate approach, which can deliver improved results, is to do a multi-staged operational training. Initial training can be provided early in the project, maybe the standard vendor course, and then a second training course is given just prior to cutover focused on refresher training and more site-specific details. This approach will increase training costs but can improve the comfort level of operators with the system and enhance retention. This strategy is also valuable because operators develop an understanding of the capabilities of the new control system early in the project enabling them to provide more insightful input for the project team.

Developing an effective training strategy can be difficult on control system migration projects when the project is completed in phases. Phased migrations often result in two active control systems being used for a period of time. This can present a number of logistics challenges in determining how to best support the operations team learning. In these cases, it is recommended that special staffing and training plans be considered to minimize the training complexity and reduce operational risks.

One common method used to address this scenario is to designate a system expert per shift and provide in-depth training on the new control system to these resources earlier in the project. As parts of the production process are cutover to the new control system, these designees are task with initially operating from the new control system. They are then expected to become trainers or "go-to" resources for the other operators on their shift as the transition to the new control system progresses. A formal training of the other operators takes place at some point nearer to a complete operational cutover. This strategy is dependent on the cutover duration being a reasonable time period and the initial cutover phase being manageable for a single operator.

Training location can be another important aspect of the operations training plan. Due to the large number of resources normally trained within the operations department, it can often be most effective to hold the training locally. However, as mentioned earlier holding training on location can often result in distractions and operators getting involved in resolving urgent issues because they are on site. For this reason, I typically recommend holding local training at an offsite facility if at all possible to minimize disruption and facilitate more focused learning.

Table 6.4 below provides a summary of recommendations for operations training.

Table 6.4. Operations training recommendations

Training Questions	Common Recommendations
Who should be trained?	Control room operators Operations supervisors and management (optional) Field operations technicians (optional)
What content should be covered?	• Display access and navigation • Process performance monitoring • How to respond to process alarms • Real-time and historical trend viewing • Changing operating parameters for control • Access, navigation, and general operational usage based on site-specific configuration (optional)
When should training occur?	A few weeks prior to cutover
Where should training take place?	Locally and offsite
How should training be executed?	Vendor course with hands on opportunities (customization can often be taught using a project resource team)

OTHERS

Outside of the core group of trainees, there are others who can be impacted by the new control system. These employees need to have a clear understanding of how to perform their job tasks when the control system migration takes place. They may not require formal training courses but may still need some basic level of training, a set of instructions, or other supporting materials. These individuals are frequently overlooked in the training plan. This can cause struggles with work completion and increase frustration levels until these individuals are educated on how to use the new control system to accomplish their job.

Project managers need to give thoughtful consideration to who might fall into this category. A plan should be developed addressing how to distribute the necessary knowledge to these job functions to enable seamless transition to the new control system. The example below highlights a case where not providing training to a work group can lead to frustrations and negative experiences with the new control system.

Example

Process engineers in your refinery frequently access graphics, trends, and control loop tuning parameters at an engineering console in the control room. Your training plan does not address providing them with information on how to do this. You have cutover the new system during a refinery outage and the refinery is starting up on the new control system. The process engineers are trying to assist the operators in troubleshooting an issue with a valve and don't know how to access the faceplates and associated parameter information. The process engineers are frustrated and their perception of the new control system is negative because of this experience.

In this example, the frustrations could have been avoided with a simple instruction sheet left at the engineering console that outlined how to access graphics, trends, and control loop tuning parameters. The importance of creating a smooth transition to the new control system cannot be understated. Developing a comprehensive training plan that addresses educating the entire organization impacted by the new control system will greatly simplify the transition and enable people to embrace the new control system.

Those impacted by a migration project are not limited to individuals directly accessing the new control system. The new control system may supply information to or require information from a third-party system or application in a different format than the existing control system. It is the project manager's responsibility to notify users of the third-party solution of any changes impacting the information they see or need to provide. Notification might include a brief training session, presentation, or video. Alternately, it may be sufficient to provide simple written instructions, directions, or other supporting documentation. Whatever method is used, project managers need to ensure that end users of third-party solutions know how to make any adjustments to how they perform their job task in response to the new control system.

SUMMARY

Successful training will form a firm foundation not only for a successful project but also for long-term operational success with the new control system. A comprehensive training plan improves cutover efficiency, simplifies system turnover, supports safe operations, and establishes a positive impression of both the project and new control system. Training plans can be complex because they require coordination and cooperation between multiple groups. Funding of training can also sometimes be complicated and needs to be clarified at the onset of a project.

The content of the training is extremely important and should deliver the needed knowledge to the right individuals. However, other factors like the timing and location of the training should not be overlooked as they also contribute to the degree of training retention achieved. Effective training plans are not only focused on primary users of the control system, but also take into account how to educate others whose job tasks might be impacted by the new control system.

Three Key Takeaways

- Training should be provided to anyone who has regular involvement with the control system as part of their job function.
- Training contributes to a control system migration project success by making the cutover and project turnover more efficient.
- Training timing and location are key factors in successful learning.

7

Progress Monitoring, Change Orders, and Reporting

Control system migration projects involve cross-departmental collaboration and require the execution of many simultaneous work activities across disciplines. Migration projects need strong project managers who understand the dynamic nature of these projects, can closely monitor progress, immediately recognize issues, and quickly adjust strategies to overcome challenges. Complete and well-defined scope, schedule, and budget documents are key enablers for successful in-progress project management activities.

Once a control system migration project is underway, project managers shift their focus from planning activities to monitoring, reporting, and change management responsibilities. These tasks are required of project managers on every project they manage, so what makes a migration project unique? Control system migration projects are a distinctive blend of both construction and software project categories. As a result, migration project managers must utilize a combination of approaches to monitor and determine progress for different aspects of the project.

Most projects clearly fall into either a construction or a software category. Granted there are very diverse projects within these categories, but the workflow and associated progress tracking within a given category is consistent. Specifically, construction projects are relatively linear in nature and progress is more easily verified. For example, to install a new piece of equipment may require pouring a new foundation and then building a civil support structure. The progress of these steps is visible and can be physically inspected and verified.

Software programming progress is less intuitive and more difficult to track. On a migration project most control system configuration activities fall under the software programming umbrella. Software programming activities are generally less linear in nature making it challenging to measure and verify progress. The example below illustrates some of these common challenges.

Example

A logic narrative must be implemented in the control system to handle the sequencing of valves on a transfer system. Due to some unique functionality, the standard function blocks will not suffice and custom function blocks must be developed. In addition, the sequence must be initiated through a start button on the transfer system graphic.

In this scenario, there are three main components of work to be completed. First is the development of the customized function blocks. Next, sequential function charts are developed using a combination of standard and customized function blocks. Finally, the transfer system graphic must be configured. The graphic is being configured by a graphics engineer while the functional logic programming is simultaneously being completed by a senior controls engineer.

The senior controls engineer starts programming the logic on Friday morning. You request a progress update on Monday morning to include in your weekly summary report to be issued later that day. The status update indicates that the custom function blocks will take a few more hours of configuration, the graphics configuration is complete, and the sequential function chart configuration has not been started. Based on the progress and remaining items, the controls engineer estimates that the team is 60% complete with the overall task, which should be completed on Wednesday afternoon.

Wednesday afternoon you check with the controls engineer and find out that he is not complete with the task as the custom function block took longer than anticipated. He has just started configuring the function chart and now estimates that he will have it completed by Thursday afternoon. As a project manager you solely relied on the assessment of the controls engineer to determine the progress and are now concerned about the inaccurate estimate.

There are two commonly encountered issues highlighted by the example above. First is the challenge of accurately assessing and verifying the status of a task while it is in progress. The second issue is the difficulty of estimating progress on tasks that are not linear in nature. One way to increase the accuracy of progress tracking for nonlinear activities like configuration or programming tasks is to use a more granular work breakdown structure. Breaking tasks into smaller sub-tasks can give you more insight into both the work involved in completing a task and an improved measurement of progress against overall task completion.

In the example above, instead of a single task for development of custom function blocks you might have included three sub-tasks to create custom function blocks for manual valve line-up exception, valve alignment verification, and tank selection. The overall goal is to have a working sequence initiated from the display. There are three activities that comprise the overall task, but they are not equal contributors. The activity of building custom function blocks is a larger percentage of the work required to complete the task than adding a start button symbol to the display. While building a custom function block represents one-third of the overall task activities, it may actually be 70% of the work. The completion percentage is a function of the effort required to perform an individual portion of the overall task relative to the whole activity. Each task should be evaluated for how it integrates with the schedule and budget allocation for the overall activity.

There are many software tools to assist project managers in handling project progress monitoring, reporting, and change order management. While tools can improve efficiency, a project manager's success is rooted in a fundamental understanding of how to accurately measure progress, report performance, and make necessary adjustments during project execution. In this chapter, we examine the unique aspects of control system migration projects from a monitoring, reporting, and change management perspective. We will define what it means to monitor a project, discuss the goals of monitoring efforts, and describe the detailed components of effective progress monitoring. We will also outline options for measuring project progress. We explain how to identify and handle changes to the project during the execution phase. Finally, we review frequency, content, and format reporting requirements.

MONITORING

Project monitoring is the on-going evaluation of progress during the execution phase against your control system migration project's established scope, schedule, budget, staffing, and quality plans. The scope, schedule, and budget documents from the definition phase of the project serve as your baseline comparison metrics. The primary goal of project monitoring is early identification of deviations from the plans. Once aware of an issue, corrective adjustments can then be made to minimize impacts to the overall project. Good project management requires you to quickly assess any deviations, understand what solutions are available, and select the optimum solution. Another goal of project monitoring is to enable proactive coordination of necessary work task timing and staffing adjustments to keep a project consistent with the defined scope, within budget, and on schedule.

Project monitoring is often assumed to be the project manager's responsibility. While it is true that the project manager has ultimate responsibility, every team member has an important role in the project monitoring process. Team members are responsible for being knowledgeable of and thoroughly understanding the scope, schedule, and budget of all work tasks for which they are involved. When they see the potential for or occurrence of deviations from plan, they should immediately notify the project manager. This will allow proactive rather than reactive project decisions.

By the simplest definition, progress metrics evaluate how actual scope, schedule, budget, staffing, and quality compare to plan or expectations. If you are doing what you said you were going to do, you are performing to plan. If on the other hand, the schedule is lagging, costs are exceeding the budget, or scope items are either not delivered or do not meet quality standards, you are underperforming. In the best case, you are trending ahead of schedule or below budget and are outperforming plan. The anecdote below illustrates the inexact nature of monitoring and tracking progress on a control system migration project.

Anecdote

My lowest point and my greatest learning moment as a project manager are one in the same. I was managing a control system migration project for a client. I had project monitoring and reporting processes in place. We were making great progress and were nearing the time to do a Factory Acceptance Test (FAT) at the control system vendor location. In the weeks leading up to the FAT, the senior controls engineer reported that he was 80% complete with the configuration.

Based on this completion percentage and the remaining scheduled items, we expected to be on schedule for the FAT so the date was finalized and the staging facilities reserved at the control system vendor's facility. Over the next few weeks, the estimates of completion percentage didn't move even though many hours were charged to the project. The initial estimates of percentage completion had been too aggressive and the remaining work was a more significant percentage of the work than the senior controls engineer had anticipated. Ultimately, we had to eliminate the step of performing a preliminary internal FAT. During the FAT, our client found numerous significant issues with the configuration that would have likely been caught and corrected in a preliminary FAT.

As a project manager I failed in multiple ways to effectively monitor project progress. I did not recognize the nonlinear nature of the configuration work that remained. This is a common mistake among project managers when any kind of software programming is involved. Another mistake was not verifying or having someone verify the work progress, which is often difficult to do for configuration or programming work. While the controls engineer on this project was very senior, he was working on the details of the project. It was my responsibility to look at the bigger picture and validate that the actual completion percentage was consistent with the work progress versus the overall task.

Just as a side note this project ended up being successful. We were able to recover from the FAT issues because of the hard work and long hours of a great project team. By the time we went to site and cutover to the new system during a planned outage, we actually completed the project on time and within budget. At the close of the project, the customer wrote us a letter of recommendation noting that although the FAT was challenging our recovery from the issues was exemplary and the overall project was a success.

The project monitoring process includes several points of comparison, which are captured in Figure 7.1 below.

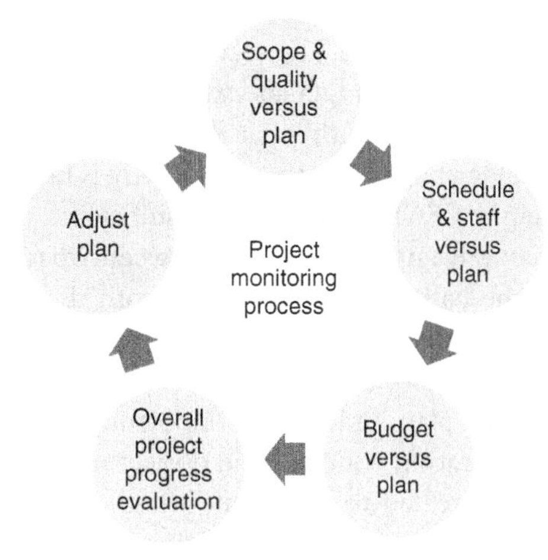

Figure 7.1. Project monitoring process.

On an ongoing basis the scope, schedule, budget, staffing, and quality should be compared to plan throughout your migration project. The overall project progress based on all of these components should then be evaluated with any necessary adjustments to plan implemented. The individual steps of the project monitoring process shown in Figure 7.1 are examined more closely in the following paragraphs.

Scope Monitoring

Comparing the actual scope being performed against the planned scope helps identify any potential change orders and is also a good way to ensure that you are producing the appropriate deliverables. Scope monitoring should verify that all aspects of the actual work being done including quantity, granularity, etc., are consistent with the original scope definition. The example below illustrates the impact of deviations from scope granularity and how scope monitoring helps identify and address potential conflicts.

Example

A controls engineer is assigned the task of creating an FAT plan. The budget and schedule call for her to develop the plan in ten hours. She develops a plan to test every detail of the new control system and submits it. The plan development took

16 hours instead of the ten hours budgeted. The plan is extremely thorough but is also 35 pages long and will take a minimum of seven days to execute. The testing period was budgeted and scheduled for four days.

There are two concerns with the FAT plan. The first is that the development time took six hours longer than budgeted. The second is that the end result is a plan that will subsequently cause additional budget and schedule overages if executed as written. There is nothing inherently wrong with what the controls engineer produced. In fact if reviewed, it might be determined that this granularity is needed for adequate testing and that the original planned FAT duration was insufficient.

The fundamental issue is an inconsistency between what has been done and the granularity included as the basis of the budget for both the development of the plan and the FAT itself. Unfortunately the additional time spent developing the FAT plan is lost. However, through scope monitoring, the project manager will be able to identify the extra detail in the FAT plan and has an opportunity to proactively address the potential FAT execution duration conflict. The project manager can evaluate the FAT plan the controls engineer created to determine whether the more granular test plan is necessary or whether the plan can be executed at a less detailed level to perform the FAT in the originally planned time and budget constraints. Without scope monitoring, the FAT duration variance would likely not be identified until the FAT was in execution leading to an extended schedule, more detailed scope, and increased budget versus plan.

Quality assurance is another integral part of scope monitoring. A project manager or a designated quality assurance manager should review project deliverables per defined quality procedures. In many cases, spot reviews of deliverables suffice to ensure that they are consistent with quality standards and defined scope requirements. Poor quality that requires re-work can be a major contributor to cost overruns on projects.

Schedule Tracking

Tracking activity timing and duration against the project schedule is also an integral part of performance monitoring. When the overall project or any portion is either ahead of schedule or behind schedule, the impact of these deviations from the planned schedule needs to be evaluated. Adjustments to the timing of future scheduled activities or to staffing plans may need to be made when schedule progress does not track according to plan. The example below highlights a situation where being ahead of schedule can actually negatively impact project costs.

Example

Your project includes a new Remote Instrument Enclosure (RIE) building that houses new marshalling cabinets. The I/O cabinets for the new control system will be located in the existing I/O room attached to the control room. The construction team has five scope items scheduled for the month.

1. Set the RIE building in place
2. Complete all RIE building grounding and electrical activities
3. Run cables from the RIE building to the I/O room
4. Install new I/O cabinets, pre-assembled with I/O cards
5. Make all field side terminations in the new I/O cabinets

The construction team completes the first three items one week earlier than scheduled. Unfortunately, the I/O cabinets required for the remaining items are not scheduled to arrive for three more days. As a result, a crew of eight construction team members continues billing to the project but cannot start their next task. They complete some preparation work and some cleanup but are not being fully utilized toward substantial project completion progress. While you could have them demobilize it is not a viable option for such a short period of time. Being aware of this situation you call the manufacturer and for a minor cost increase are able to get the shipping of the cabinets expedited so they arrive two days early. This allows the construction team to get back to full productivity minimizing their inefficiency.

The above example indicates that any deviations from plan whether favorable or not can impact the project. Frequent monitoring of critical schedule milestones enables early identification of predecessor and successor relationships that can go off-plan creating a cascade of schedule issues. By identifying these and other schedule conflicts early, necessary adjustments can be made.

Schedule deviations not only impact the project, but can also extend to external work groups. For instance, if the cutover is planned for a major maintenance turnaround but the project is behind schedule, the turnaround timing might be impacted. Likewise, external events like a postponed turnaround where cutovers were planned can have an impact on the project schedule. Staffing plans also need to be validated in conjunction with the schedule. Are the resources assigned to a given task aligned to meet the schedule completion targets? Monitoring the scheduled and upcoming activities, knowing the interdependent relationship of scheduled tasks, and being aware of external activities that can affect the project are important project management coordination responsibilities.

Budget Evaluation

Comparing project spending against the budget is probably the most common way to identify off-plan events. When spending exceeds budget and progress is not exceeding the plan, it is a red flag that there is an issue. It is imperative that project managers know when any aspect of the project is deviating from the budget. There are a wide variety of causes for budget deviations and among the most common are as follows:

- Inefficient or unproductive resources
- Performing expanded scope without change orders
- Inaccurate estimating and insufficient budgets.

When resources are billing to the project but not making progress consistent with the plan, the project can quickly exceed the budget. Common reasons for poor resource productivity include rework due to quality issues, misunderstanding of expectations, and inexperience or a lack of knowledge. Resource inefficiencies can also be caused by schedule issues. For example, the construction team cannot complete a work task because the design team has not issued the design packages for a particular area.

Work performed beyond the defined project scope without documented change orders also frequently contributes to significant budget overages. Many times, team members eager to please and wanting to add value take on work that is not part of the original scope. Often they do not compare tasks to the original scope and never even recognize they are doing work that is in addition to what was intended or planned.

Estimating is an inexact science and as a result, it often contributes to budget overages as well. It is vital that all team members understand the assumptions that are used as the basis of the budget. If the assumptions are clearly communicated and the project is efficiently executed consistent with those assumptions but exceeds the budget, then there is a good chance that the original estimate was not adequate. Unfortunately when this is the case, there are few proactive steps you can take to correct the budget overages unless you have some contingency money you can apply. Alternately, you can introduce change orders to cover the overages citing estimate deficiencies as the root cause.

Broad terms for progress status such as performing to plan or underperforming are insufficient as meaningful budget evaluation metrics.

The quantification of a financial metric is generally considered the best and most easily validated representation of project progress. The most minimalistic budget metric is a calculation of the percentage of the budget spent. In an ideal and linear world, this would not only be a budget status, but also a reasonable progress metric. For example, if the budget is accurate and you have spent 30% of the budget, then you have completed 30% of the project.

In reality budget spent is only a spending metric that provides a very rough approximation of project progress. This metric does not take into account actual progress toward completing a work task or alignment with schedule milestones. Figure 7.2 below shows an example spreadsheet with a basic percent of budget spent calculation.

Gathered From Financial Reports

Calculated From Spent Divided By Budget

Project Tasks	Budget	Spent	Line Item % Spending
Configure I/O Points	$10,000	$7,500	75%
Configure Control System Hardware	$5,000	$5,000	100%
Configure Complex Logic	$15,000	$20,000	133%
Configure Graphics	$20,000	$15,000	75%
Configuration Activity Totals	**$50,000**	**$47,500**	**95%**

Figure 7.2. Basic percentage of budget spent spreadsheet.

Figure 7.2 includes line items for high-level control system configuration activities. Each line item task in the spreadsheet also includes the budget, the amount spent to date, and the percentage spent, which is calculated by dividing the amount spent by the budget. From these items, we can determine what our spending progress is versus the budget. The overall budget spending for all configuration activities is at 95% of budget. The complex logic configuration task is 33% over budget, the I/O point and graphics configuration tasks are 25% under budget and the hardware configuration line item is 100% of budget.

We are unable to determine from these budget spending numbers actual progress toward completion. Is the complex logic configuration task that is 133% of budget actually complete? Of the two tasks that are 25% of budget is one complete and the other incomplete? What we know is that we have spent 95% of budget so if we have more than 5% of configuration work tasks remaining to complete, then we are trending over budget.

Tracking spending can give early insight into potential project issues. For example, if you know that spending for a particular task is 50% of budget, but the task progress is not at 50% then you might expect it to trend over. You might consider making a proactive adjustment to avoid the overage. Comparing actual spending against budget also enables you to track spending rates, which can be used to estimate future project spending. This can be a useful way to validate the project progress. However, to determine overall project progress, we need to do a more complete analysis than spending versus budget.

Overall Progress Calculations

Beyond the basic spending versus budget evaluation, more advanced progress calculations provide valuable insight into project performance, deviations from plan, and projected variances impacting future activities. The most holistic measure of progress includes technical, cost, and schedule components. There are numerous degrees of implementation of advanced progress calculations, which need to be matched to your specific project size and complexity. You do not want to implement a progress tracking approach in your project that is cumbersome and significantly raises project management cost if it is not necessary.

There are many different methodologies for tracking project progress, and we could not possibly cover all of them here in the detail required to adequately represent the concepts. We will cover some basic conceptual details and identify some approaches that can be used for lean implementations of progress tracking on control system migration projects.

A relatively basic need is a measurement of technical progress versus goal for the money spent in a specified timeframe. We will call this concept basic earned value analysis. It has been my experience that basic earned value analysis is adequate for progress tracking on most control system migration projects. Basic earned value analysis requires tracking three elements including the budgeted cost of work scheduled (BCWS), actual cost of work performed (ACWP), and the budgeted cost of work performed (BCWP). Each of these terms is defined and better explained below.

Budgeted Cost of Work Scheduled (BCWS) is the budgeted cost of a task per a specific evaluation date.

Actual Cost of Work Performed (ACWP) is the actual cost required to complete all or a portion of a task per a specific evaluation date.

Budgeted Cost of Work Performed (BCWP) is the value earned from the work performed by a specific evaluation date.

The example below further clarifies the definitions of these terms and their relationships to one another.

> **Example**
>
> Your control system migration project contains a line item to complete configuration of five overview graphics over a five-day period. Each graphic is expected to take one day to complete at a cost of $800 per graphic. The work is started on Monday morning and is planned for completion Friday afternoon.
>
> At the end of the day on Wednesday, you perform a basic earned value analysis. The BCWS in this scenario is $2400 calculated by multiplying the budgeted cost of $800 per day by three days. The ACWP is $1800 because a junior resource configured the graphics at a bill rate of $75 per hour rather than senior resource at the $100 per hour budgeted. Your BCWP is $2000 because the third graphic is not complete and your progress is estimated at 50% rather than the 60% planned. The BCWP in this example is calculated by multiplying the budgeted cost of the overall task $4000 by the estimated completion percentage of 50%.

What meaningful information can we gain from the example above? First, we can identify that both technical progress and schedule are lagging slightly behind the plan. However, the costs are actually under budget. Even though the schedule is lagging by 10%, the favorable cost variance of the lower bill rate of the junior resource covers that gap and will enable a recovery to plan without exceeding the budget.

The above basic earned value analysis is relatively straightforward, but the most commonly misunderstood aspect of this approach is how to determine the BCWP or earned value. There are a number of ways to estimate or calculate the earned value of a task. The first is to manually enter an estimated percent complete based on the time spent versus task duration combined with the progress made. This is a somewhat subjective approach that should factor in the nonlinear nature of the task. For example, if you have completed three out of the six steps required for a task, but you know that the last three steps represent 75% of the task duration, then your estimated completion percentage is not 50% but instead 25%.

A second approach is to use the estimated physical percent complete. This is a measure of progress based on the number of component steps completed without considering the factor of time duration or nonlinearity. For example, if you have a task to configure ten graphics, then when you have completed five graphics your percent complete is 50% and your subsequent earned value or BCWP is 50% of the budget for that line item. It is possible that the last five graphics might take half of the development time of the first five graphics, but this is not considered as a factor in the physical percent complete estimation.

An alternate approach to the estimating methods outlined above is to establish pre-defined earning rules for your tasks. These earning rules do not provide the most accurate updates on task that are in progress but bring consistency to the calculation and simplify the process reducing the amount of time invested in determining BCWP. As an example, a rule might be established that no completion credit is given until a task is complete so that your earned value is either 0% or 100%. Just as easily a rule might be established that when a task is started 50% completion credit can be claimed with the remaining 50% credited when the task is completed. These rules work especially well on projects that are simple, short in duration or have a very granular work breakdown.

Any of these approaches can be used to estimate completion percent. Once completion percentage is determined, it is multiplied by the task budget to determine the BCWP or earned value. Factors that help determine the best approach for your project include the level of detailed analysis desired, the time you want to invest in tracking progress, the complexity of your project, and the granularity of your work breakdown structure.

An even more advanced methodology for tracking project progress is the earned value management (EVM) concept. There are various levels of implementation of EVM that extend the basic earned value analysis concepts, enable you to calculate cost and schedule variances, and predict future project spending and schedule progress. There are definite benefits to using the earned value management methodology, but it also requires additional time and budget for managing your project unless you already have established tools to utilize. We will not go into further detail on the EVM concept here. There are many books, courses, and available information on the internet to introduce this concept in more detail if you prefer using a more advanced approach.

Adjusting Plans

The control system migration project that executes 100% to plan is a rarity. Making adjustments to project scopes, schedules, budgets, and staffing plans are a normal part of project management responsibilities. The key to making successful adjustments is to know which variables can be modified when project variances are identified. The fixed relationships between scope, schedule, and budget introduced earlier as the project management triangle combined with project staffing are the typical control variables for a project. When a single one of these areas is a problem, the others become control variables or potential constraints that help define the appropriate response as shown in the example below.

Example

Your progress report reveals that the construction activities have fallen behind the schedule while cost is tracking to the plan. You immediately evaluate the options for getting the project schedule back on plan.

1. Extend the schedule for these activities if it does not adversely affect other project activities. This would not require additional resources but would likely increase project costs associated with these construction activities.
2. Add resources to help the construction team catch up to the schedule. This would raise cost as unplanned resources would bill to the project and the availability of resources, their ramp up time, etc. can all be factors to consider.
3. Reduce the scope of construction. Maybe there are activities that were extras that can be pulled from the project? This is usually not the case and reducing scope is seldom a viable option once a project is in the execution phase.
4. Shift future work activities assigned to this construction team to others. Maybe another construction team is available, which can handle some of the future task planned for this construction team? This would allow the existing team to complete their current work on an extended schedule assuming that it does not impact the start of future work activities.

On your project the scope is fixed, and because you are planning cutover activities for a pending turnaround, there is limited flexibility with the schedule. You are aware of some contingency money that you have, which could be used so that there is some budget flexibility. You are also aware that the construction team has several resources working on some other noncritical path items of the project. You choose the first option and add resources from another part of the project to help recover the schedule. You also increase the budget using some of your contingency to pay for the additional headcount charging to these construction tasks.

When multiple constraints exist often times your options are limited and the situation can be complex to overcome. It is incumbent upon the project manager to balance the various aspects of the project, recognize the options, and select options which optimize project performance based on priorities and constraints. Beyond minor project adjustments, change orders are often necessary to handle major changes that occur as part of the project.

CHANGE ORDER MANAGEMENT

Change order management is a natural extension of project monitoring activities. Changes will occur during control system migration projects that expand the scope, alter the schedule, and increase the budget. Some of these changes are conscious deviations from or adjustments to the project that are discussed and planned in advance. Other changes are much more subtle and can go unrecognized until after they have derailed the scope, schedule, or budget. Project monitoring is an essential exercise to help identify these more subtle changes and convert them from reactive responses to proactive and managed changes.

At the foundation of good change order management is a well-informed and knowledgeable team. The project manager has ultimate responsibility for managing project changes. However, it is the responsibility of every team member to have enough familiarity with the project in their respective discipline area to recognize potential changes and bring them to the attention of the project manager. While there are certainly some differences in the formal change order management processes among companies, the conceptual workflow of handling potential change orders is the same, as shown in Figure 7.3 below.

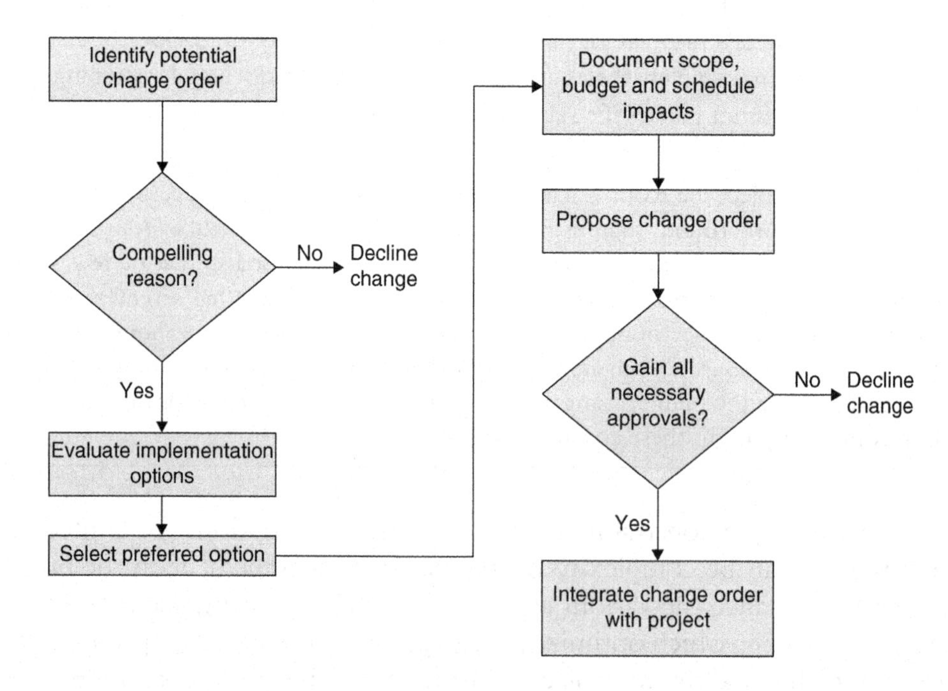

Figure 7.3. Common workflow for handling potential change orders.

When a potential change is first identified, it is often documented in a request for information (RFI). RFIs are used to gather information and further investigate a potential change to the project's scope, schedule, or budget. RFIs are not change orders but a prelude to a change order and generally alert those involved in the project that a change is being considered.

A first step after identification of a potential change is to determine whether there is a compelling reason for the change. The compelling reason may be a significant benefit to the company with a favorable cost-to-benefit ratio, or it may be that not executing the proposed change will delay or prevent other project activities that are part of the existing project scope. Changes must generally be justified based on their technical, financial, safety, maintenance, or operational value merit.

When it is determined that a compelling reason exists to move forward with considering the change, then often times different implementation options need to be evaluated. Option evaluations can be technical or might be related to the implementation timing or the magnitude of the scope change. The example below outlines a case where several options might be considered for a potential change order before determining the best approach.

Example

During the course of executing your control system migration project, the design and construction teams discover some issues with instrument signal cables from a field junction box. The junction box has four different multi-pair cables and the issues are with only one of the cables affecting approximately sixty instruments. It is considered a risk to the integrity of the new control system installation and you make the decision to propose a change order.

One option you consider is replacing the specific multi-pair cable with the shielding issue. Alternately, since all cables exiting the junction box are routed in the same cable tray back to the marshalling room, you consider a second option of replacing all four of the multi-pair cables. All of the cables are over 20 years old and updating all of them might alleviate potential future maintenance issues.

The costs and schedule impacts of replacing all four multi-pair cables are minimal above and beyond replacement of the single multi-pair cable and will eliminate future field cabling concerns associated with that entire junction box. As a result, you recommend replacing all of the cables exiting the junction box and routing into the marshalling room as the best option.

Once a preferred option is selected, the project team thoroughly documents the proposed change with respect to scope, schedule, and budget. In general, the proposed change needs to be fully documented as a standalone project although part of the scope may be the integration with the overall migration project. Once this documentation is complete, a formal change order can be created using the end user company's change order format. The change order is then submitted for review and approval by all necessary parties.

If a change order is not approved, then no further action is required. When a change order is approved, the project manager is then responsible for integrating the change order into all aspects of the project, including scope, schedule and budget documents, staffing plans and project monitoring and reporting activities. It is often valuable to track change orders independently as "mini-scopes." This can allow project managers to better measure the impact of changes to the project. For example, if there were five change orders to the project and three of them tracked over budget, then it will provide insight into both the completeness of the initial scope and the accuracy of budgeting in the change order process. While these metrics may not be immediately useful in the project, they can be valuable in the post-project analysis. Outside of this, approved change orders become a part of the project just like any other work tasks.

PROJECT REPORTING

Project reporting is the mechanism for communicating meaningful project information to team members, company management, and other interested parties at some regular frequency during the project execution phase. The three key elements of reporting are frequency, content, and format. The specific reporting approach is largely defined by company requirements, available tools and information as well as management expectations.

Reporting should be done at frequent enough intervals to keep appropriate individuals and workgroups adequately informed of the project progress. However, reporting too frequently can unnecessarily increase costs and add workload to key resources for little value. Alignment with available financial reports, the size of the project, and the weekly project spending rate are all important factors to consider when determining the appropriate reporting frequency.

In order to have accurate and current metrics in a project progress report, the distribution cycle must be properly aligned with financial reporting. For example, if a project manager receives time sheet summary reports on Wednesday covering the preceding week, then issuing a project progress report on Tuesdays would be off-cycle. In this situation, the information in the project progress report would be stale or inaccurate because it would be based on the last time sheet report received, which would actually be for work executed two weeks prior. To improve accuracy and maximize value, plan progress report distribution for shortly after receipt of new labor and financial information as long as it is timely. Reporting too far back in history because you waited on information defeats the purpose.

The size of the project also helps determine the appropriate reporting frequency. Companies define project thresholds differently, but I suggest using staff numbers as a rule-of-thumb measure of project size. I use the following definitions as my personnel project thresholds, but they can easily be adjusted for your organizational needs. Small projects are those with a maximum total engineering, procurement, and construction staff of less than 25 resources.

Mid-size projects can be defined as projects with staffing between 25 and 50 while large projects have headcounts exceeding 50 resources.

Smaller projects can often be tracked on a bi-weekly or even monthly basis and adequately keep everyone informed of the project progress. Mid-size and large projects have enough complexity that they often benefit from more frequent reporting. Reporting on a weekly basis even in times of lower activity on mid-size or large projects is often valuable if for no other reason than to force spending analysis by the project manager.

Another variable to consider in establishing reporting frequency is the weekly project spending rate. When weekly project spending is either a large nominal amount of money or a high percentage of the overall budget, any off-plan events can have an enormous impact and need to be identified as soon as possible. The example below illustrates the importance of frequent assessments on high spending rate projects.

Example

Your weekly project spending is $50,000 for a project with a total budget of $1 million. The expected progress toward completion in each week is 5% and you have chosen to report on a monthly basis. You are two months into the project and complete a report showing the project is tracking to plan.

The week following your reporting is a poor progress week and though spending is at plan, progress is 2.5% or only half of plan. When you get to the reporting period three weeks later, you identify the issue and recognize that you are now trending $25,000 over budget. If this had been caught immediately, you could have potentially made adjustments to offset the lack of progress. Since that time you have spent an additional $150,000 and have little room for adjusting activities at this point to correct for the slow progress week that occurred a month earlier.

It is extremely valuable to have a higher frequency of reporting when the spending and budget size warrant. This enables you to identify variances up front as they occur and gives you the flexibility to make necessary adjustments to address the issues and compensate if necessary in other areas of the projection the other hand, reporting too frequently can burn valuable resource time and drive project cost higher. It is not uncommon to adjust reporting frequency during various stages of a control system migration project. For example, project reporting may be less frequent during the early design phase of the project but then increase during critical milestone periods of construction or cutover.

It is important for project managers to establish and gain acceptance of report frequency, content, and layout at the beginning of a project. Creating new reports or being required to continuously answer questions during project execution because reports have not provided the needed information is both frustrating and inefficient. Reporting, when approached properly, is a natural extension of regular project communications, monitoring, and tracking activities, and should not require significant additional time investments.

All team members have the responsibility of providing accurate and timely information to include in reports. Delayed or missing information is one of the biggest contributors to reporting errors. For example, late time sheets that miss a reporting period can prevent an accurate depiction of project spending. Report formats vary greatly among companies based on software tools and financial reporting systems being used. Reports can be delivered via email, through file sharing applications, or using business intelligence software programs. The delivery approach should maximize accessibility to the intended audience. This is important because you want to make it easy for those on the distribution list to use the report and gather needed data.

Reports should be organized to highlight any areas of concern or required action items. Reports should also leverage the existing information available in project documents such as financial reports, budget spreadsheets, and schedules. Excerpts or screenshots from these documents can also be embedded into your reports or they can be used as attachments to provide supporting data or details. The most basic report is a summary progress report, which is brief and provides a general overview of the project progress and status.

An example simple text report is shown in Figure 7.4 below.

Bi-weekly control system migration report
March 4–15, 2013

Project completion metric
Overall earned versus spent is 67%. Progress percentage this period is 8%.

Key milestones completed this period
I&E design
- Developed detailed cutover plan
- Updated I/O index with new addressing assignments

Controls engineering
- Configured distillation area graphics
- Developed transfer system function charts
- Developed factory acceptance test plan

Construction
- Installed power supplies for controllers
- Installed new power panel for control room

Key activities in progress
- Installing cables between marshalling cabinets and new I/O cabinets
- Configuring logic for catalyst area batch sequences
- Detailed cutover planning

Page 1 of 2

Key milestones planned for next period
I&E design
- Continue detailed cutover plan development
- Finalize last set of loop sheet drawing updates

Controls engineering
- Configure catalyst area graphics
- Configure logic for reactor controls
- Develop interface details for advanced control application
- Developed site acceptance test plan

Construction
- Install controllers in I/O cabinets
- Install workstation in temporary location in control room

Outstanding action items or issues
- Operations review and approval of distillation area graphics needed by end of next week
- Need to identify ways to speed daily work permits currently taking 1 hour past shift change
- Need management review and approval of pending change order #3

Page 2 of 2

Figure 7.4. Basic project management progress report.

There are five fundamental elements to include in basic reports:

- Estimated progress
- Key milestones completed
- Key activities in-progress
- Key activities planned
- Open action items or issues needing resolution

Estimated progress is a completion percentage represented in a quantified metric (e.g., earned value). The key milestones completed section identifies the specific tasks completed since the last report at whatever granularity you determine to be appropriate. The key activities in-progress section summarizes the tasks that the team is currently working on and provides estimated completion percentages on these tasks when available.

The key activities planned section outlines those activities that are expected to be started or worked before the next report. This is a particularly valuable section in that it communicates activities that might need to be planned for in the work responsibilities of others. For example, if you are going to start moving new workstation equipment into the control room, this is the information that the operations department would want to know and the report can be a reminder to operations management of the planned work.

The final section covers open action items or issues needing resolution. This section highlights issues that are limiting progress, key decisions that are waiting on inputs or approvals, and other areas impacting the project, which require support from others. I did not include a commercial section in the base report because the value of this section depends on the nature of your project team and contract terms. It does not necessarily have to be a standalone section and might be rolled into the open action items or issues section. However if you are a vendor, system integrator or EPC company, you definitely want to include a commercial section, which highlights any contract, invoice or payment milestones, or issues. It can be a great tool for raising awareness on commercial items, which can often take time to resolve.

More advanced progress reporting includes additional detail but also requires more time to compile and assemble information. In addition to the basic reporting content described above provided in a more granular detail, common additions to advanced reports include those listed below.

- RFI log
- Change order log
- Deliverables status list

- Summary status of all invoicing to date
- Separate subcontractor activities
- Critical upcoming schedule milestone dates
- Resource loading
- Planned spending curves

One area that is often focused on with more granularity for project reports is representing completion progress in a more holistic manner potentially including projections of future progress performance. Figure 7.5 below shows a more detailed earned value progress calculation that can be used in more advanced project reports.

BWCP Is Calculated from Multiplying the Budget By the Estimated % Complete

Project Categories	Budget	Spent	% Spent	Estimated % Complete	Earned Value in Dollars (BWCP)
Project Management	$20,000	$10,000	50%	47%	$9,400
I&E Design	$75,000	$70,000	93%	98%	$73,500
Controls Engineering Design	$25,000	$20,000	80%	96%	$24,000
System Configuration and Integration	$75,000	$35,000	47%	52%	$39,000
FAT, SAT and Training	$40,000	$10,000	25%	21%	$8,400
IE&C Construction	$1,00,000	$50,000	50%	50%	$50,000
Cutover	$50,000	$0	0%	2%	$1,000
Final Project Documentation & Closeout	$15,000	$0	0%	0%	$0
Configuration Activity Totals	**$4,00,000**	**$1,95,000**	**49%**	**51%**	**$2,05,300**

Figure 7.5. Earned value summary information.

Alternately, you might want to use a bar chart to show the monthly spending and earned value in comparison to your budget as illustrated in Figure 7.6 below. This allows you to look at monthly performance to gain insight into your monthly actual versus expected performance over the course of the project.

In some cases, you might want to include a line graph indicating how the overall earned value is tracking versus the cumulative budget. This can be conveyed in graphical format as shown in Figure 7.7 below.

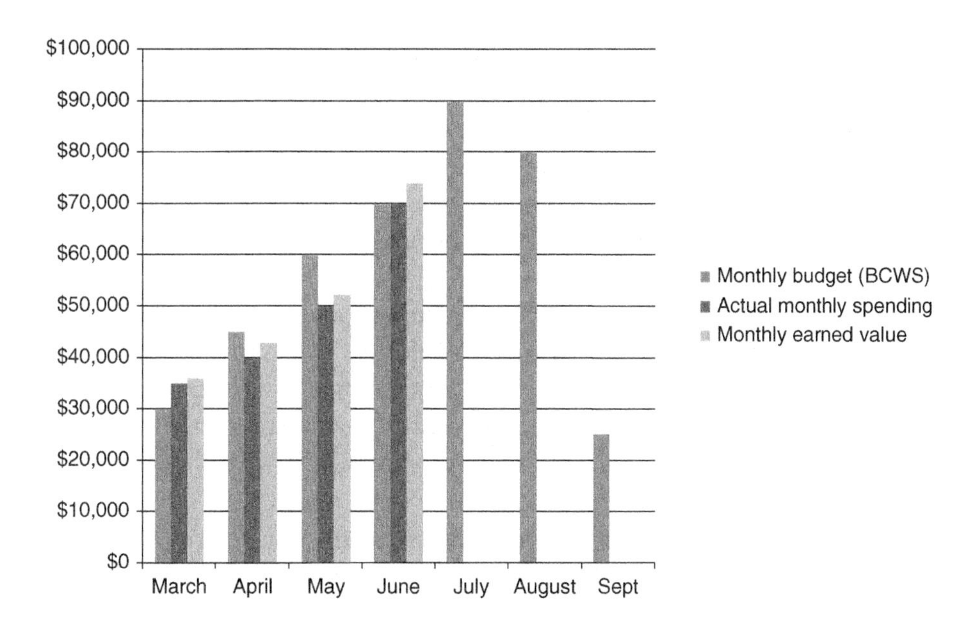

Figure 7.6. Monthly spending and earned value versus budget chart.

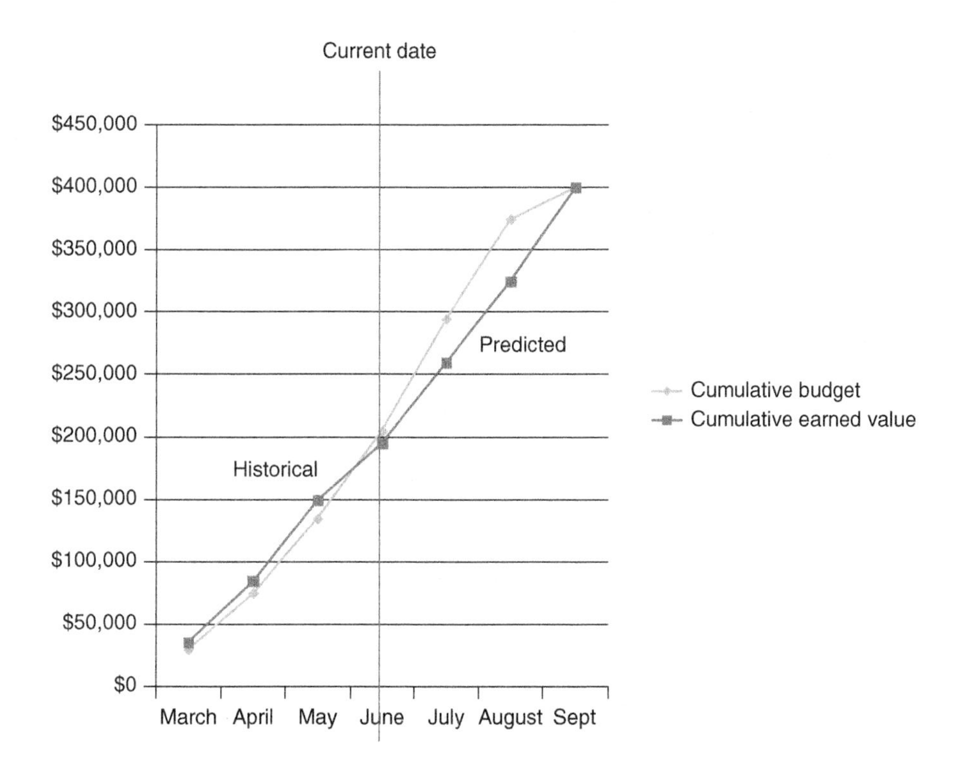

Figure 7.7. Overall earned value versus cumulative spending curves.

The goal of this graph is to show whether the earned value of the overall project is trending ahead or behind the planned budget based on the calendar schedule. It can also be used to predict performance through the remaining months of the project. The current date is indicated by the vertical red line on the graph, which in this case is June. Everything to the left is the historical calculated earned value. The remaining months to the right are predictions of earned value performance.

The cumulative budget curve depicts the expected monthly budget based on the planned work. You may or may not choose to make adjustments to the predicted future curves based on your projects historical actual to budgeted spending performance. For example, if you have been trending behind on the budget 30% each month, then you may want to adjust your projected budget profile to more realistically align with actual performance.

The biggest challenge is predicting future trajectories for the earned value. There are multiple ways to estimate this. The simplest method but least accurate is to use an equal calendar distribution approach. Using this method, you calculate the remaining value to be earned on the project, which is the difference between the final earned value that should equal the overall project budget and the current earned value to date. You then divide the remaining value to be earned by the remaining number of months in the project schedule. This will give you an amount which you then use as the estimated monthly earned value for each of the remaining months. Equal monthly earned value is seldom an accurate representation but can be a quick means to build out the future prediction of earned value in the graph.

An alternate method is to look at the percent of work budgeted or planned each month through the remainder of the project. This number is relatively easy to determine by using the budget and schedule tools available. You can then multiply the monthly budgeted percentage of whole by the remaining value to be earned to get an estimated monthly earned value. An example of this method of calculation is provided below.

Example

Referencing Figure 7.7 the monthly budgets for the remaining three months of the project are $100,000, $70,000, and $30,000, respectively. This would correspond to planned monthly budgets of 50% in July, 35% in August, and 15% in September. Assuming that the remaining value to earn is $205,000 then the estimated monthly earned values calculated by multiplying this amount by the monthly budget percentages would be $102,500 for July, $71,750 for August, and $30,750 in September.

The approach outlined in the example above may take a bit more time to calculate, particularly at the early stages of a project, but also tends to be a more accurate baseline. However, if you are trending way behind or ahead on your earned value, then you probably want to make some adjustments to these numbers by using a bias factor based on your historical project performance information.

There are numerous ways to depict meaningful progress analysis in reports. The main priorities should be to provide consistent reports with fixed assumptions and common progress calculations. Do not shift between using a monthly spending and earned value chart one reporting period and projected spending and earned value graph the next. These inconsistencies will make it confusing for the report users. Also, if you predict future earned value using a percentage-based calculation, then do not shift to a calendar-based calculation for earned value predictions later in the project.

Reporting should not be just a required exercise that project managers have to do but instead should add value to the project. Accurate, meaningful reports benefit the project manager, project team members, other workgroups involved in the project and company management. The exercise of developing regular reports can often lead to the discovery of project issues that would not have otherwise been identified in a timely manner.

SUMMARY

Project monitoring, change order management, and progress reporting are important project management responsibilities that keep your control system migration project on scope, within budget, and on schedule. Accurate monitoring and reporting of control system migration project progress begins with a strong project definition rooted in the scope, schedule, and budget documents established in the definition phase of the project.

Project monitoring is the frequent review of project progress during execution against metrics established in the definition phase. It involves the entire project team understanding the scope and accurately estimating progress toward completion of key activities. Project monitoring can help identify deviations from plan early. This enables proactive management of these deviations whether through scope, schedule, budget, and staffing adjustments, or via change orders.

Even in the best planned and most well-defined control system migration projects, changes can occur, which expand the scope, alter the schedule, or increase the budget. Identifying these changes is the responsibility of every team member. There is a workflow to handling change orders that begins with their identification and requires thoughtful consideration to the compelling reason behind their necessity, the evaluation of implementation options, and the selection of the most appropriate option within the context of the overall project.

Project reporting is the mechanism for communicating meaningful project information to project team members, company management, and other interested parties during project execution. The frequency, format, and content of reports as well as their delivery to the target audience are all relevant factors that should be considered in determining project reporting details. Project reports should not require large amounts of time outside of normal project activities but instead should be designed to leverage available information without require significant additional time investment.

Project monitoring, change order management, and progress reporting require an engaged project manager and knowledgeable project team. Evaluating actual project performance versus plan and developing a meaningful progress metric will assist you in staying on course to meet expectations established at the beginning of the project.

Three Key Takeaways

- Project monitoring activities enable early identification of deviations from plan and allow proactive coordination of needed schedule, scope, budget, and staffing adjustments.
- All team members have a responsibility to be knowledgeable of the scope, schedule, and budget associated with their work activities, recognize occurrences of off-plan events, and communicate variances to the project manager.
- Effective project reports are distributed at a reasonable frequency, communicate meaningful status and progress updates to the appropriate individuals, and leverage existing project information to minimize additional time investment.

8

High-Risk Areas

Risk management involves the identification, analysis, and control of risks to avoid or minimize undesirable consequences. Every project has risks associated with it. In the process industries most project risks have safety, financial, schedule, or operational impacts. Within the context of the project, the risk include failing to meet the scope, schedule, or budget goals. There are numerous risk assessment and risk management philosophies from simplistic, informal discussions to complex, well-documented risk matrix processes. Independent of the risk evaluation methodology you use, the outcome should be a firm understanding of your project risks so that you can evaluate their probability of occurrence, estimate the impact or severity if they do occur, and assess their manageability. Using this information you can then formulate a strategy to best address these risks.

Every task of a control system migration project risks jeopardizing the scope, schedule, and budget if not executed correctly or managed properly. Areas of the project that are not defined with enough detail during design and engineering are especially prone to unforeseen issues during the construction phase. On migration projects there are several areas that can be at particularly high risk to expand your scope, extend your schedule, and exceed your project budget. Table 8.1 below lists common high-risk areas for control system migration projects and provides descriptions of each risk with corresponding mitigation suggestions.

Once you have developed a risk mitigation plan, it is wise to regularly evaluate high-risk areas to ensure that your plan is working. The remainder of this chapter will discuss each of the high-risk areas from Table 8.1 below in more detail. We will examine the risks, explore how they can impact your project, and review specific steps that can help lower their likelihood of occurrence.

Table 8.1. Summary of high-risk areas

Project Area	Risk Description	Risk Mitigation Suggestions
Graphics	Poor graphics planning and design leading to significant changes during the review process.	• Develop a comprehensive graphics design guide during the FEL or as part of early project design and engineering efforts. • Gather operations input early in the graphics development process.
Graphics	A poorly defined graphics review and approval process.	• Document the responsibilities of all parties in the graphics review and approval process and manage to the process.
Graphics	Unrecognized complexities in the old control system graphics.	• Early in the project, validate assumptions regarding the dynamic functions in the old control system graphics.
Third-party system or application communications	Inconsistent use and application of communications protocols and standards.	• Thoroughly research the options and specific setup requirements as part of early engineering efforts. • Schedule a meeting at the beginning of the project with all parties to work out communication details up front.
Third-party system or application communications	Coordination of multiple vendors to troubleshoot and resolve issues.	• Involve vendors early in the engineering process to gather their recommendations and identify any potential inconsistencies. • Schedule a meeting at the beginning of the project with all parties to work out communication details up front.
Staffing changes	Changes in project resources that cause additional ramp-up time and lack of familiarity with project details.	• Emphasize a transition plan and communication exchange when resource changes occur.
Poor teamwork	Lack of communication or responsiveness among departments and team members.	• Schedule regular communication and status update meetings with all parties. • Involve all departments impacted by the project in planning activities focusing on identifying and addressing any concerns.

(*Continued*)

Table 8.1. (*Continued*)

Project Area	Risk Description	Risk Mitigation Suggestions
Unforeseen logic complexity	Logic programming is more complex than expected and requires additional programming time.	• Ensure logic narratives developed during the FEL or early in the project are thorough. • Assign experienced staff to logic programming tasks.
Field construction obstacles	Combinations of physical space limitations and operational issues that impede field construction activities.	• Coordinate constructability reviews between the design teams and the construction manager. • Plan regular meetings with management to review the project progress and status.
Cutover Details	Lack of detailed planning that prevents smooth cutover execution.	• Ensure that the cutover plan is detailed and thorough. • Coordinate detailed cutover plan reviews between the design teams and the construction manager.
Cutover Details	Execution mistakes that impact operations.	• Schedule daily cutover meetings to emphasize mistake-free execution over speed of execution.

GRAPHICS

Graphics are often a headache for project managers on control system migration projects. There are a number of opportunities for graphics design and configuration to easily get off track impacting scope, schedule, and budget. One common risk with graphics is significant changes during the review process. This problem often originates when there is a lack of good planning and design or limited upfront involvement by operations. Large numbers of graphics in most manufacturing operations mean design changes and rework affecting them have the potential to have a substantial financial and schedule impact on the project.

The risks associated with poor graphics design can be mitigated. A comprehensive graphics design guide is one of the key tools to reduce graphics design variability. This guide should address all major areas of graphics design and provide a roadmap for graphics configuration when the project gets underway. The operations department should be asked to provide input into this design guide as a way of involving them early and better understanding their specific design requirements.

Along these same lines, another way to mitigate graphics variability is including operations in the graphic design and building process. This can be accomplished in a number of ways. One method is to have select operations technicians build the graphics. This is preferable but in many companies this is not practical because it often requires a full-time commitment of operations personnel for some period of time. An alternate method which does not require as much dedicated time is to have a small group of operations technicians become consultants for the graphics development team. These technicians contribute to the graphics design process by answering questions for and making suggestions to the graphics design team. They also represent operations in the graphics review and approval process.

Whatever method you decide to use, it is extremely important to integrate operations into the graphics development process from the beginning of the project. Rather than building a large number of graphics in bulk, it is recommended to build a few graphics that represent the different types and complexities, and then get approval for these as the templates for the rest. This can be an effective methodology in resolving any issues early in the process. There will be disagreements even among different operators about the best approach to certain graphics design areas because each person will have individual preferences. However, early operations involvement gives you the opportunity to proactively resolve conflicts or issues before investing a significant amount of time building graphics which then require rework.

A second area of difficulty with graphics arises when there is a poorly defined review and approval process. There is a tendency to want to continually make tweaks to graphics, especially as features with the new system are discovered. The best way to eliminate a continuous graphics change cycle is to clearly define who reviews and approves changes, how many reviews will take place, and when in the graphics development cycle they take place.

Various stakeholders will have diverse perspectives and different preferences regarding the best approach to any number of graphics design areas. Without an established review process, multiple people will request changes from the first release of graphics for review through the startup and transition to operations. A firm review and approval process combined with a good graphics design guide will force departments and personnel to align their expectations early in a project. When this does not occur, the graphics can become a source of frustration that can impact the perception of the overall migration project. The anecdote below recalls a situation where a lack of operational input to the graphics design process adversely affected the control system migration project causing unnecessary additional investment.

Anecdote

I managed a project, which involved rebuilding a large number of graphics in a new control system. We were told to use the existing control system graphics as the general design basis. The client project manager was a relatively young process engineer. One of the features he appreciated in the new system was the ability to have 3-D graphics. As part of the graphics design guide, he insisted on building 3-D process piping that dynamically filled with a certain color based on the process flow.

Our team made every effort to convince him that this approach was going to be difficult to manage and many operators would not like operating from these screens. Unfortunately, he insisted on using this approach and so we developed the graphics based on this design philosophy. It was time-consuming and not nearly as straightforward as the control system vendor had convinced the end user client it would be.

We went through an established review and approval process with the client project manager taking the lead role as approver. After loading the approved version of the graphics, we immediately started receiving feedback from operations that they were unhappy with the graphics, particularly the 3D process piping, which they found to be a distraction. When the operations manager saw the graphics, he was also unhappy with the design and expressed concerns about the usability.

A few weeks after the original project was completed, we were asked to submit a proposal to revise the graphics. The plant subsequently hired us to make additional changes to the process graphics so that they would be easier for operations to use. The graphics including the revisions eventually cost the plant a significant amount more than they should have because there was a lack of alignment among key end-user stakeholders during the initial graphics design process.

Another risk area that can significantly impact scope, schedule, and budget is unrecognized complexities in the old control system graphics. This is applicable when graphics for the new control system are not being completely redesigned but instead use the existing graphics as a partial or complete design basis. Estimating the scope, budget, and schedule to configure graphics in the new control system is typically made based on evaluating a sample or subset of graphics from the old control system. There is an inherent risk in this approach of missing key dynamic or scripting elements on individual graphics. Graphics often have functionality that is not obvious when looking at a static view.

In many older control system graphics, dynamic functionality required scripting or logic functionality programming. For instance, in many older systems to have a pump change colors with run and stop status changes requires coding or scripting in some programming language. This scripting can sometimes be extensive and complex. While much of this functionality is now easier to configure in newer control systems, the back engineering process can be time consuming. Back engineering involves reviewing the scripting in older graphics to develop a complete understanding so that the dynamic functionality can be accurately built into the new graphics. Scripting or additional graphics functionality that is not identified during the scoping stage of the project may lead to more graphics configuration effort required than originally planned. The anecdote below outlines the challenges that scripting can cause when estimating graphics development time.

Anecdote

I managed a project that involved upgrading operator consoles and building roughly 300 graphics in the new control system operating environment. There was no conversion program so the graphics were being developed in the new system from scratch and the client wanted to replicate the old graphics with the exception of adding some new dynamic capabilities.

Our proposal was developed based on the information provided by the customer as well as a brief spot check of graphics from the old control system. Unfortunately, during this process, we failed to recognize that a number of the graphics in the old system involved large amounts of dynamic scripting. It took tremendous effort to review and understand the functionality of all of this scripting and then build it in the new graphics. As a result, we exceeded the estimated budget and missed the original scheduled completion date by several weeks.

Mitigating risks associated with not identifying graphics scripting and dynamic functionality can be a challenge, especially when there are a large number of graphics. End user companies may not always be aware of the amount of scripting involved in their own graphics. Particularly, on older graphics that have been in place for a number of years and have not recently been revised or modified, the complexity behind the functionality may not be understood. Discussing graphics with the control room operators and observing them as they operate the unit can often be a quick way to help identify unique or nonstandard functionality on graphics in the older control system.

One potential mitigation step is to include a work task in the project to perform a detailed graphics review in the early stages of the migration engineering effort so that the complexity of each graphic can be fully understood and documented. Definitions for what are considered low-, medium-, and high-complexity graphics need to be developed as well as an estimate for configuration time for each classification. You can then evaluate individual graphics placing them in the appropriate complexity categories based on the definitions. When all graphics are categorized, you can use your estimate for each classification multiplied by the quantity of graphics in that category to develop an estimate, which you can then cross-check against your original graphics configuration estimate identifying any discrepancies. You can then determine what adjustments might need to be made if large variances exist.

THIRD-PARTY SYSTEMS OR APPLICATION COMMUNICATIONS

Those most experienced with control system migration projects will often tell you that one of the biggest challenges is to establish interfaces to or integration with third-party systems and applications, such as serial devices, historians, etc. These communications should not be difficult to configure as in many cases standards are used. However, the number of different interfacing and communications protocols available combined with the variability of how standards are implemented between different vendors can make this a very complex part of the project as the example below explains.

Example

Consider the OLE for Process Control standard, which was first established in the mid-nineties. The acronym has since been redefined as Open Platform Communications (OPC). The premise was a standard communication methodology between Windows™-based software applications and control systems that would promote open connectivity. A variety of OPC standards have been developed through the years such as OPC Data Access (DA), OPC Historical Data Access (HDA), and OPC Unified Architecture (UA). While the evolution of these has represented progress in key areas, not every vendor has the same capability or strategy to adopt these different versions into their products in a timely manner. As a result, the standardization effort still has variability dependent upon what OPC standard the vendor is using and the degree to which they have implemented the applicable OPC standard. This results in inconsistent engineering effort requirements to implement an OPC solution between different vendor's solutions.

The example above regarding the OPC standards illustrates the uncertainty associated with communications configuration. This same challenge is true about many other connectivity approaches as well. For this reason, it is important to be very clear about what connectivity standards will be used including the version of the standard and how the particular vendors involved from both the control system side and the third-party application implement the standard.

To reduce the risk of connectivity and communications adversely affecting the project, you will want to thoroughly research the communications options and specific setup requirements for each third-party connection as part of early engineering efforts. Establishing cable pinouts and setting up driver communications can often take significant upfront time before a single point is working. Once those issues are resolved, the mapping of remaining points is relatively routine.

It is also valuable to involve third-party vendors early in the engineering process. They are an important resource to help answer questions and guide you through the configuration process. One suggestion is to schedule a conference call or meeting early in the project with both the control system vendor and the third-party system or application vendor to establish expectations, discuss any potential issues, and work out communications details. Establishing these relationships early can also help with later coordination of vendor communications if issues do arise and troubleshooting support is needed.

STAFFING CHANGES

In an ideal world the project team including resources from the end user, control system vendor, and EPC provider is constant throughout the control system migration project. Unfortunately, staffing changes occur with regular frequency on projects. It may be personnel changes with any of the parties involved in the project, all of which have the potential to negatively impact the control system migration. Each time that resources are changed, there is a ramp-up time or learning curve involved to become familiar with the project and establish a context for the work that the resource is involved in. Particularly, when those with in-depth project knowledge are changed, the new resources can shift the entire logic and approach behind key strategies in the project. The example below highlights a circumstance where staffing changes can adversely impact a control system migration project.

Example

You hired a system integration company for the EPC services on your control system migration project. The lead control system engineer worked on the FEL and is very knowledgeable of the logic in the system having documented it in narratives. He also has configuration experience with both the old control system and the new control system. He has completed the initial hardware and system administration configuration activities.

Configuration of I/O, control loops, and logic is planned over the next two months. He announces the Friday before this work is to begin that he has found a great job opportunity elsewhere and resigns from the system integration company. The system integration company assigns one of their other senior control engineers to the project. She has experience configuring the new control system but has not worked with the existing control system before.

> The new control engineer spends her first two weeks gaining familiarity with the project and the old control system. While this is necessary very little earned progress is being made toward tasks completions for the budget being spent. When she starts configuring the logic, she repeatedly encounters challenges with understanding the narrative details and must evaluate the code in the older control system with which she has little familiarity. As a result of this single resource change, the controls engineering activities fall several weeks behind and the original budget, which was in part based on resource efficiency due to the familiarity of the resource with the code, has now been exceeded.

The example above illustrates the impact that resource changes have on a project. While it might be argued that in this example the system integrator did not provide a person with equivalent knowledge as a replacement because of her lack of familiarity with the older system that is an unrealistic expectation. Resources are seldom interchangeable and the knowledge and skill sets of people do vary. This is one of the reasons that as you consider how to resource your control system migration you will want to evaluate the number of employees a potential service provider has with knowledge applicable to your project.

One strategy to help address staff changes is to double staff in key positions. For example, you might want to have two senior control system engineers working on the project as opposed to one. This approach can provide you with a safety net in the case of staffing changes. The downside is that this can increase project costs and can lead to inconsistencies in the implementation if they are not wholly aligned.

When resource changes occur, mitigating the impact requires good communication. As staffing changes are identified, the project manager should establish a transition plan whenever possible. This plan should provide some level of overlap and information exchange. If this is not possible directly between the original and replacement resource, then it is incumbent upon the project manager or a designee to take this role. The resource that is exiting the project should be asked to provide a summary status of activities. It is also helpful to have a list of questions that the exiting resource can provide written responses to as part of the information exchange so that it can be used as a reference document for the new resource.

In these situations, the project manager also has the responsibility for minimizing the ramp-up time of the new resources. This may require other project resources to spend additional time with the new resource explaining design logic, answering questions, or otherwise adding context to the project. Resource changes during projects do occur and the best way to mitigate the impact is by establishing good communication and spending some initial time educating the resource to increase their project knowledge quickly.

POOR TEAMWORK

Poor teamwork has contributed to the failure of many control system migration projects. In general, the characteristics of poor teamwork include little or no communication and a lack of responsiveness. Poor teamwork causes project inefficiencies often in the form of schedule delays or rework. The project team itself must successfully work together across disciplines to avoid issues and to resolve challenges when they do occur. For example, the design and engineering team might need to gather input from the construction team regarding how to best route a cable tray in a particular high elevation area. Open communication and the free exchange of ideas between these different discipline groups on the team can head off conflicts between design and constructability.

The need for good teamwork is not isolated to the project team. The cooperation of various departments in the company with the project team is just as essential. If the operations department is indifferent to the project, or worse yet adversarial to the project, this will impact project execution. If the I&E group doesn't feel they are being considered in the design process, they may not be as supportive to the construction team. Involving departments impacted by the project from the beginning is a key to being successful as evidenced by the anecdote below.

Anecdote

While working as the lead control systems engineer in a plant I received a call from an EPC company that had been hired to do the FEL work on all plant projects. The engineer introduced himself and then explained that he was working on a project at the plant assigned to another project manager within our company. The project required a number of analog and digital points in the control system. The engineer asked me for some control system addressing assignments, which I provided. He immediately told me that my I/O assignments wouldn't work. Due to the location of the field devices and the junction box he was planning to use, all of his signals were coming from a specific field junction box and since he was using a single home run cable he needed to land them all in a single marshalling cabinet.

I explained that we had just completed a multi-year project to separate the digital and analog signals in separate marshalling and I/O cabinets and considered it a plant design philosophy. He remained adamant that I was wrong and told me that he would get the project manager involved. In further discussions, I was able to determine that he had never been to our plant and was trying to do all of his engineering and design remotely.

Once I talked with the project manager and explained the issues, he quickly contacted the engineer to revise his approach. The assignments were handled per our plant philosophy. However, that particular EPC company lost all credibility with me. It is not acceptable to put the priorities of a project over the long-term end user company philosophies without getting buy-in and support from those it impacts. When a project team's scope is complete, the resources that use and maintain the system over the long term should be able to embrace it.

Improving teamwork is often facilitated through regular communications between all parties. Scheduling regular project status and communication meetings that include representatives from all team disciplines and from the departments impacted by the project can help achieve this. These meetings promote project involvement from all affected parties, give everyone an opportunity to voice concerns, and provide the opportunity to resolve potential conflicts in a managed way.

UNFORESEEN LOGIC COMPLEXITY

Accurately estimating the amount of programming time logic configuration will require can be challenging. There are many variables that impact programming time including the individual control system, the complexity of the logic, the skills of the programmer, and the program approach taken. As a result, it is not unusual to underestimate the complex logic configuration effort for control system migration projects.

In some cases, the logic is being developed solely based on the required functionality without considering how it was programmed in the old system. This makes sense in many cases because many of the logic functions that required special programming in older control systems can now be accommodated through combinations of standard function blocks in newer control systems. In these situations, programming time is often easier to estimate.

In other cases, it may be necessary to extract logic from the old system and back engineer it to determine the intent behind all details of this logic which can be a time-consuming task. The functionality of this logic must then be created in the new control system using a combination of function blocks and customized programming. Accurately estimating how long this process will take for all logic programming required on your project is difficult. To establish a budget general assumptions are made and rules of thumb are used. There is an inherent risk of deviation from budget and schedule associated with logic configuration as a result of unrecognized complexity during the estimating process. Thorough logic narratives are an important part of reducing this risk and should be developed during the FEL or early in the project design and engineering effort.

Developing a logic complexity table can be a useful tool in estimating and tracking logic configuration time. Similar to graphics, you want to develop definitions for low-, medium-, and high-complexity logic. You can then classify the logic required on your migration project in these categories. Table 8.2 below provides a representative example.

Table 8.2. Example logic complexity table

Name	Description	Complexity	Estimate (Hrs.)
Unit1ReactA	Unit 1 Reactor A headspace control	High	12
CatPump101	Catalyst pump sequencing	Low	4
Unit1SD	Unit 1 interlock shutdown sequence	Med	8

On migration projects, tasks such as hardware setup or I/O configuration can commonly be completed by less experienced controls engineering resources without impacting quality or schedule. However, this is not the case with logic configuration. Due to the variability and inefficiencies in logic programming, it is recommended that experienced controls staff familiar with the control system be assigned logic configuration responsibilities to help reduce risks of budget and schedule issues.

FIELD CONSTRUCTION OBSTACLES

While there are obstacles in every discipline area of a control system migration project, this is especially true about the construction scope. Well-conceived designs on paper do not always translate to practical construction in the field. There are a number of factors that can make field construction difficult. One of the most common is physical space limitations. Marshalling and I/O rooms in older facilities generally have minimal space remaining. New buildings or room additions can be costly so many control system migration projects use the existing room and often use a staged transition and replacement strategy. As a result, construction teams are often adding equipment in areas with limited physical room creating tight work spaces and limiting the number of construction resources that can work on a particular task at a given time. This can create construction inefficiencies and extend the schedule of these work activities.

While it is impossible to foresee all of the potential construction issues that can arise, there are ways to minimize the occurrence of issues. Engaging the construction manager or other construction team members in the design process can often improve a project's design and reduce deviations from design in the field. An important tool to employ on your project to align construction and design efforts is a constructability review. Construction team representatives and the design team representatives walk the project together in the field reviewing the design and identifying any potential construction issues. These reviews may sometimes result in design changes, but these changes prior to construction are normally less costly than when they are identified and must be resolved during the construction phase.

Another challenge to construction efforts is operational issues. The priority will always be on operating the facility. This means that when events occur that impact process operations, the project construction efforts may be delayed or stopped altogether. The delays can range from a slower permitting process to clearing a particular area or location of construction personnel to enable company maintenance activities to take place. The anecdote below outlines a situation where circumstances in the plant caused our project team to lose efficiency before we identified a workable solution.

Anecdote

Our construction team had moved new cabinets into a combination marshalling and I/O room. These new cabinets would be used to relocate various I/O from the old cabinets and then begin to replace the internals of those cabinets as they were emptied. These older cabinets as they were freed up would then be reused for the new control system. At the end of the project a few of the excess older cabinets would be removed from the space actually creating more available floor and work space than in the original room layout.

We had six construction I&E technicians working in the room wiring the new cabinets and preparing them for cutover activities. The plant had an equipment issue that caused a shutdown. It required a number of the end user company's I&E resources to be in that same marshalling and I/O room troubleshooting the issue. Our team was asked to stop work temporarily and leave the marshalling and I/O room until the plant issues were fixed.

While we were allowed to return a few hours later after the issue had been corrected, the end user company's I&E manager asked that in the future we limit our construction team to two technicians in the room at a given time during the day shift. The plant felt that because of the tight work space and number of construction personnel it limited their own I&E technicians access to the room. While we understood the issue, it changed our staffing plan substantially and threatened to impact our project schedule. As a work around, we gained approval from the customer and began working a crew at night with four technicians that in effect maintained the same staffing level and schedule.

Unfortunately, many of the operational issues and events that can impact the project cannot be planned for or avoided. If they are occurring frequently or significantly impacting the project, regular meetings with management may be helpful. These meetings with management can be used to review the project status, plan for any scheduled operational events, and raise awareness of issues that impact the project.

CUTOVER DETAILS

Cutovers are an opportune time for control system migration projects to be derailed. We will examine the challenges and best practices associated with cutovers in the next chapter. Here we want to highlight the most common risk areas and what steps can be taken to reduce the risks. There is a tremendous amount of work by every discipline required to contribute to a successful cutover. Many issues that arise in cutovers could be avoided with detailed design and thorough planning. The example below emphasizes the importance of considering all of the details during the design process such as constructability to avoid cutover complications.

Example

The control system migration design team has developed a full cutover package with drawings and detailed plans. One of the steps in the cutover process for your project is relocating I/O from one of the existing marshalling cabinets and distributing among two new marshalling cabinets that have been added in the same room. The existing field cables will be reused for this plan.

During the cutover, the construction team begins relocating the cables and determines that the cable length is several feet short of reaching one of the new marshalling cabinet locations. The cutover window is limited and the team does not have time to run a new cable from the field. The solution agreed upon is for the construction team to add a junction box to terminate the existing cable and then install a new cable from the junction box to the new marshalling cabinet. While this solves the problem, it adds a junction location that was not planned for or desired by the client and requires that a large number of drawings be updated during the as-built process to reflect this change.

The design teams and field construction teams should walk through the cutover process ahead of the actual cutover verifying that the cutover plan has considered all of the practical issues that might arise during the cutover such as the one outlined in the example. There are many other issues that can surface without thorough detailed planning as well. For instance, the sequencing of I/O change over for hot cutovers is critical. If several I/O points or loops are part of a common logic function, then the order of how and when those are moved over to the new control system becomes critical. If the design does not take into account this sequencing, controls may not work properly as they are partially in the new system and partially in the old system. Often times issues such as these fall through the cracks without good communication between the controls engineering function and the design teams.

Another high-risk area for potential issues is cutover execution mistakes, which can impact operations. Particularly on hot cutovers wiring missteps have the potential to cause operational shutdowns as indicated in the anecdote below.

Anecdote

I managed a control room consolidation project that also involved a complete conversion of the operator workstation to a new platform. The project required the installation of new control room furniture in a staged approach as well as rewiring all of the existing hardwired control room devices, some of which were unit emergency shutdown switches. The cutovers were being done hot. The first unit was cutover without incident and was operational with the new workstations and furniture.

During the cutover of the next unit, the construction I&E team was in the process of reterminating the wiring for one of the emergency shutdown switches when the bare wire was touched against a piece of metal in the furniture while it was being reterminated. This caused a short and shut down the unit. All the project work was stopped and we held an incident investigation. The biggest contributor to the incident was determined to be the sense of urgency that the team felt which lead to carelessness.

The anecdote above illustrates the fragile nature of cutover execution work when field wiring is involved. These types of events sometimes happen and are difficult to completely prevent. On many cutovers, a sense of urgency is emphasized. One way to reduce incidents associated with execution mistakes is to have morning cutover meetings where a priority is put on safe and incident-free cutover versus speed of cutover. This won't necessarily prevent all incidents, but it can help raise the awareness of the team to take their time and focus on mistake-free execution. It is important that you have considered these types of risks in developing your cutover plan and have outlined the appropriate responses when issues do occur.

SUMMARY

Risks exist on every project and control system migration projects are no exception. It is important to

- identify risks up front
- evaluate the likelihood of a given risk occurring
- understand what the impact will be if it occurs
- determine how much control or manageability you have to prevent it.

Once you understand each of these elements relative to your project risks, you can put together a complete plan for managing them. Typical high-risk areas on control system migration projects include graphics, communications with third-party systems and applications, staffing changes, poor teamwork, unforeseen logic complexity, field construction obstacles, and insufficient cutover design details. Focusing on these specific areas during the detailed design and engineering can often avoid issues during project execution.

At the most basic level control system migration projects typically have risks that impact the project's scope, schedule, and budget. However, even more importantly some control system migration project risks can also affect safety and process operations. By taking proactive steps to mitigate these high-risk areas, they can be sufficiently addressed so that they are simply another step in your migration project.

Three Key Takeaways

- Every task of a control system migration project risks jeopardizing the scope, schedule, and budget if not executed correctly or managed properly, but there are common high-risk areas that may need additional attention.
- Developing a comprehensive graphics design guide, involving key operational stakeholders early in the design phase, and establishing a clear graphics review process are essential to keeping graphics design efforts within budget and on schedule.
- Establishing communications with third-party systems or applications is often not straightforward even when standards are used.

9

Cutovers

In the previous chapter, we identified cutovers as one of several control system migration project high-risk areas. The planning and management of cutovers is such a core contributor to control system migration success that it warrants a deeper examination in this dedicated chapter. Mistakes planning, managing, or executing the cutover can cause a migration project to fail even if your project is effectively designed, properly configured, and successfully constructed.

Safety needs to be a primary focus throughout your control system migration project. However, the cutover is a time to highlight and emphasize job safety to the highest degree. When a cutover takes place you typically have the largest group of project personnel in the operating unit being exposed to the most process hazards. The cutover team should have a keen awareness of their surroundings as they execute cutover tasks. Job safety analysis in advance of cutover activities, daily safety meetings, and regular field safety audits can all help contribute to a safe and incident free cutover.

Cutovers are complex because they are an amalgamation of all aspects of the project that must be executed in a cohesive and structured manner. Strong leadership is vital in coordinating cutover activities and pace, as well as communicating plans and progress. I encourage you to approach and manage the cutover as a mini-project within the overall migration project giving due attention to both the design and execution details.

The key elements for achieving cutover success are identified in Figure 9.1 below.

In the remainder of this chapter, we will explain and outline important aspects of each of these elements in more detail. We will review common cutover challenges specific to each element providing suggestions on ways to avoid them. Finally, we examine cutover management best practices addressing planning, coordination, communication, and reporting functions.

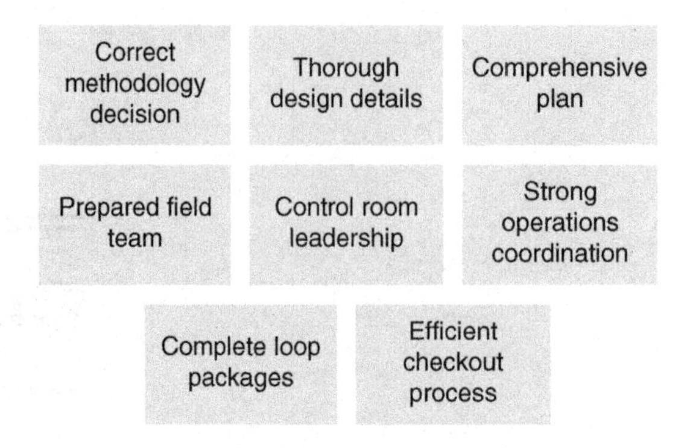

Figure 9.1. Key elements of cutover success.

CORRECT METHODOLOGY DECISION

Determining what cutover methodology will work best for your control system migration project is a major decision that helps define other aspects of the overall project. The initial decision of whether to manage and execute the project as a single, comprehensive control system migration, or employ a phased implementation approach is heavily influenced by how you anticipate executing your cutover. When we talk about phased migration in this context, it can have a couple of different meanings. Phased migration can mean different system components are replaced in phases. For instance, the controller hardware is replaced in one phase and the consoles or operator workstations in another phase. Alternately, phased migrations can mean converting portions of the facility, such as particular areas or units in phases.

The example below shows how the cutover methodology can influence the overall migration project strategy.

Example

You are the engineer at a chemical plant and are trying to determine whether to approach your migration as a single project or in phases. The plant has three operating units. The total I/O count of the current control system is 9,000 points relatively equally balanced between the units. All of the units are operated out of a single, centralized control room. Adding outages for individual units to facilitate cutover activities is not financially viable.

Due to the nature of the process, hot cutovers will not be possible and cutover activities will have to take place during planned turnarounds. Normally, the plant has major turnarounds every two years that last a few weeks. A phased migration would reduce the number of points to be cutover on a given outage to a much more manageable level. However, if you choose to handle each operating unit's migration individually, the last unit will be cutover an estimated four years after the first unit is converted. This is impractical from both an operational and cost of ownership perspective because it would require operating from and maintaining two control systems for those four years. As a result, you determine that the best project approach is to have a single, comprehensive migration of the plant's control system.

The primary migration project decision related to cutover methodology is whether to use hot cutovers, cold cutovers, or a combination of both. Hot cutovers are done loop by loop while the process continues to operate. This approach requires both the old and new control systems operate in parallel for a period of time until the cutover is complete. Cold cutovers are done as a single, complete conversion from one control system to another during process downtime with the process restarted on the new control system. Alternately, a combined approach can be employed where certain points are cutover hot while others may require an outage.

The decision of whether to cutover your control system cold, hot, or using a combination of the two is driven by a number of considerations as identified below:

- Nature and complexity of process operations
- I/O point count
- Logic and configuration complexity
- Cutover cost
- Staffing requirements and availability
- Safety implications
- Outage constraints
- Physical control room environment

Normally you prioritize these factors and then make a decision that optimizes the balance of these elements. However, any one of these factors can be a major determinant that overrides the other considerations. For example, if your process operations are highly dangerous and numerous shutdowns are programmed into your control system logic, then you will likely be forced to employ a cold cutover methodology regardless of other implications, such as higher cost or added complexity.

Some of the key strengths and weaknesses of each cutover methodology are identified in Table 9.1 below.

Table 9.1. Cutover methodology comparison

Cutover Methodology	Advantages	Disadvantages
Hot	• Does not require outage • Enables flexibility in cutover pace • Provides operations the opportunity for gradual learning of new control system • Fewer distractions during cutover activities • Easier to troubleshoot issues • Fallback option to old control system	• Risk unplanned shut down of process operations • Requirement to operate and maintain multiple control systems for cutover duration • Requires sequencing of point, loop, logic and graphics cutovers • Requires more physical real-estate • Longer duration
Cold	• Does not require simultaneous operation and maintenance of multiple control systems • Eliminates risk of unplanned outage or downtime • Does not require sequencing of point, loop, logic, and graphics cutovers • Short cutover duration • Minimizes physical real-estate requirements	• No flexibility with cutover pace • Operators must startup on new control system • Can be conflicts with other turnaround activities • Challenging to troubleshoot issues • No fallback plan, old system is eliminated
A Combination of Hot and Cold	• Provides substantial cutover flexibility • Enables reduced cold cutover activities • Provides operations the opportunity for gradual learning of new control system • Fallback option to old control system	• Requires most complex planning for sequencing of point, loop, logic and graphics cutovers • Risk unplanned shut down of process operations during hot cutover activities • Requirement to operate and maintain multiple control systems for cutover duration • Requires more physical real-estate • Long duration

You might notice that cost is not listed in Table 9.1 as a strength or weakness for any of the approaches. I frequently hear claims that there is a higher costs associated with a hot cutover. In my experience, migration cutover costs are driven by the specifics of your particular project more than the particular methodology you employ. Cost should definitely be considered as a factor in determining your cutover methodology, but it should not be assumed that cost with any single option is higher. The way that you calculate the cost will be part of the equation. For instance, if you use the cold cutover approach do you consider lost production part of the project cost? The profit gain by keeping a unit in operation is typically an order of magnitude higher than project labor costs. You must analyze the cost associated with the various methodologies applied to your particular project before drawing a conclusion.

If you select a hot cutover methodology, the cutover duration is longer, but the number of resources required for the cutover activities, particularly on the field side, might be reduced. However, hot cutover costs are increased by the need to operate and maintain multiple control systems. Hot cutovers do allow you to start taking advantage of the new system earlier, which can have return-on-investment (ROI) implications. This is an especially important consideration if outage opportunities for cold cutovers are infrequent and can delay conversion to the new system for a significant period of time.

A cold cutover is a shorter duration but often requires more staff to support the cutover activities because of the limited outage window. You must perform a large amount of work in a small amount of time. Depending on the number of points to convert, overtime cost can be a substantial financial burden with this methodology. Additionally, any delays in the outage timing for the cutover might require you to demobilize your field team for some period of time or have them billing to the project without optimal utilization.

Ultimately there is no single cutover methodology that is universally the best choice. The cutover methodology must align with the specific requirements and needs of your project. Give thoughtful consideration to which cutover approach will minimize disruption to process operations, maximize the learning experience with the new control system, and work best within your project's scope, schedule, and budget.

THOROUGH DESIGN DETAILS

Much like the overall project, the foundation of successful cutover efforts is established with thorough and detailed design. There are a number of critical aspects to the design effort including an accurate, comprehensive, and clear cutover I/O database, as well as detailed and verified wiring instructions and transitional equipment location, installation, and demolition details.

Your project I/O index is the basis of your cutover I/O database. Depending on how you establish your project I/O index, it may include all of the fields necessary for cutover. However, this is often not the case and some additional field designations might need to be added. A complete I/O cutover list should include any applicable fields from Table 9.2 below:

Table 9.2. Suggested I/O cutover list fields

New control system tag	Old control system tag	Description
Signal type	Range or scale	Engineering units
Other links	Old system I/O addressing information (card, slot, etc.)	New control system I/O addressing
Old marshalling panel termination details	New marshalling panel termination details	Old junction box termination details
New junction box termination details	Loop drawing number	Notes or remarks

You may not need all of these fields, but you should make sure that any relevant fields are in your database. For example, if you are using the same tag naming convention, which is generally preferred unless there is a compelling advantage to renaming, you will only have a single tag name field in your database as opposed to a field for the old control system tag and another for the new control system tag.

Most of the fields in Table 9.2 are self-explanatory. The "Other Links" field may not be as intuitive. This field identifies any pertinent connections or relationships for a given I/O signal that need to be considered in cutover planning. These might include utilization of the I/O in software connections, such as advanced control logic, interlocks where the I/O point is used or for multiple uses of the field wiring like a signal splitter. Utilization of points on process graphics can also be a valuable field to track. The example below illustrates the importance of documenting I/O links during detailed design to avoid operational issues during cutover execution.

Example

Your I/O cutover database includes flow control loop FIC-100 with an input FI-100 and an output FY-100. The loop is part of an advanced logic algorithm configured in the control system, which also includes LIC-101 and TI-115. You decide to use your existing I/O index and the loop sheets rather than developing an I/O cutover database. Your existing I/O index does not include a field to designate I/O links.

You develop your hot cutover plans and identify FIC-100 to be cutover on the first day. You have scheduled LIC-101 and TI-115, which are located in a different I/O cabinet to be migrated on the fourth day. As the cutover begins, your team converts FIC-100. The operator immediately recognizes control issues with the loop. A quick investigation reveals that the advanced logic is not working because LIC-101 and TI-115 have not been moved to the new system yet. FIC-100 is moved back to the old system and rescheduled for cutover on the fourth day with the other associated points comprising the advanced logic.

The example above highlights the importance of having a field within the database for clear linking of I/O point relationships. Missing the identification and documentation of subtle I/O links is one of the most common contributors to cutover issues. If you are performing a hot cutover and overlook one of these relationships, it can result in operational upsets or downtime. While documentation such as loop sheets may identify these relationships, during the chaos of cutover it is often difficult to ensure that all of these links are recognized by all parties involved in the cutover process. It can be extremely valuable to have a single database that includes relational link information as a reference.

One of the best ways to make sure you capture the links associated with an I/O point is by holding I/O cutover review meetings. The purpose of these meetings is to complete a point by point review of the I/O list with the I&E and controls design and engineering teams. This gives everyone an opportunity to identify and document any unique requirements for each point in the database. This process may be timeconsuming, but it will ensure a complete database that can be used as a primary and centralized source of information during the cutover.

Performing a detailed field wiring verification is another key cutover design responsibility. It is not enough to rely on drawings and documentation to develop wiring instructions. The designer should field verify all of the wiring is routed and connected per the existing documentation. It is not uncommon for wiring terminations to be misaligned with documentation, which can create a cascade effect of issues if not identified during the design efforts. In addition, it is common to find physical wiring issues that can create havoc on a migration cutover as highlighted in the example below.

Example

You are executing a hot cutover for your control system migration project. The first step of the cutover is to de-terminate, relocate, and reterminate a small 16-pair cable in an existing cabinet. As the cutover begins, the field cutover team quickly calls your attention to convoluted wiring in the existing cabinet. The wiring in the cabinet is tangled in several places. Because many of the cables in this cabinet provide signals and control for critical operational functions, the risk is deemed too high to try to untangle the wiring. The design team did not recognize this nested wiring as a constructability issue. The cutover plan for the first day is abandoned and a new strategy must be developed.

Cable routing issues also frequently create challenges during cutovers. Particularly, if existing cables will be reused and routed to other locations, the cable lengths and routing paths must be physically verified to avoid surprises during the cutover. For instance, when routing under raised floors, obstacles such as floor supports, cable trays, other cables, and miscellaneous equipment can impede preferred cable routes and create constraints that shorten usable cable lengths.

In many cases operator workstations, engineering workstations, and possibly marshalling or I/O cabinets must be installed in temporary locations during hot cutovers. Accurate design details regarding where to locate these and how to provide power and wiring to them are required. This also involves physical verification of factors, such as cable lengths, cable routing obstacles, and physical space considerations. For instance, in crowded rooms with limited real-estate something as simple as door swing radius on a cabinet can make a temporary location infeasible. Design consideration should also be given to minimizing rework and avoiding logistical issues when equipment is relocated to its final location at the end of the cutover.

The example below shows how a lack of attention to these design details can create unnecessary challenges during the cutover.

Example

Your cutover involves both hot and cold cutover activities. In order to effectively operate during the hot cutover activities two operator workstations for the new control system are being installed in temporary locations in the control room. The construction team installs both workstations in their temporary locations per the design.

When the construction team lifts the raised floor panels to pull the power cables to the workstations, they discover that the temporary location sits directly on top of a wiring and equipment panel for the plant paging system. The design engineer had not lifted the raised floor to verify the availability of free space below the temporary workstation locations. Because the paging system is a critical part of the plant communications infrastructure, the designed temporary location is not acceptable. The workstations must be moved to the other side of the room requiring ad hoc design changes be made.

The example above emphasizes the importance of physical field verification as part of detailed design. There are many hidden challenges that affect constructability, and it is imperative to identify those during design efforts rather than cutover execution to avoid costly fixes and schedule delays. Employing sound design and engineering practices will help you avoid many common cutover issues. A majority of these issues, as discussed in this section, are about physical field constraints that are only identified as a result of thorough design efforts.

COMPREHENSIVE PLAN

The cutover plan is a critical document, which is the equivalent of a combination of the scope, schedule, and staffing plan for the cutover. Effective cutover plans expect the unexpected and have contingency processes built into them. For example, if you are doing a hot cutover and a point is moved to the new system and isn't working, how will you handle that? Anticipating issues and accounting for those scenarios in your cutover planning efforts will reduce the impact when off plan events do occur.

At a minimum your cutover plan should address:

- The scope of cutover work to be done
- Which resources are responsible for what portions of the work
- When the work is planned
- How the work is sequenced and executed

Clear instructions on the cutover execution and loop checkout process will help define the scope of work for the cutover. There are variations in the I/O checkout process for different point types so each one needs to be clearly defined. Part of detailing the plan also involves identifying what staff functions are responsible for various parts of the cutover. For example, who will be responsible for initiating the lifting of wires on the field side? Will the coordination be driven by the end user company operator, the field I&E cutover team, the control room operator, or the control room cutover manager? All team members as well as operations need to have a clear understanding of roles and responsibilities during cutover.

The project manager or a designee is responsible for putting together a detailed schedule of cutover activities. Cutovers should be organized and grouped in manageable blocks. Groupings might be by a particular process area or by a particular cutover day. For example, you may want to designate those points planned for cutover on the first day to the group D1, the second day points to D2, etc. You can then sort or filter your spreadsheets by these groupings to easily create your daily I/O cutover list. Figure 9.2 below shows two example days from a sample cutover planning spreadsheet. Note that in order to keep the example simple not all fields from the I/O cutover database are included.

DAY 1 I/O Cutover Plan

Number	Tag	Description	Point Type	New Control System I/O Location	I/O Links
1	FI-097	Outgoing Effluent Flow	Analog Input	1A-1-1	Used in Environmental Calculation Logic
2	AI-095	Cooling Water PH	Analog Input	1A-1-2	Used in Environmental Calculation Logic
3	FAL-095	Cooling Water Pump Low Flow	Digital Input	2D-1-1	
4	TI-090	Ambient Temperature	Analog Input	1A-1-3	Displayed on Utilities Overview Graphic
5	LI-091	Wastewater Level Indicator	Analog Input	1A-1-4	

DAY 2 I/O Cutover Plan

Number	Tag	Description	Point Type	New Control System I/O Location	I/O Links
1	FI-100	Solvent Flow To RX 1	Analog Input	1A-2-1	
2	AI-102	RX 1 Off-Gas Analysis	Analog Input	1A-2-2	
3	PI-100	RX1 Pressure Indication	Analog Input	1A-2-3	Input to PIC-100
4	PY-101	RX1 Pressure Control Valve	Analog Output	1A-5-1	Output from PIC-100
5	TI-100	RX1 Solvent Temperature	Analog Input	1A-2-4	
6	LI-100	RX1 Level Indiciation	Analog Input	1A-2-5	
7	HS-104	Catalyst Pump Handswitch	Digital Output	2D-5-1	Used in Catalyst Sequencing Batch Logic
8	LAH-100	RX1 High Level Alarm	Digital Input	2D-1-2	
9	TI-101	RX1 Cooling Jacket Temp.	Analog Input	1A-2-6	

Figure 9.2. Cutover planning spreadsheet.

Another major component of the cutover planning detail is to understand and account for sequencing of points, loops, logic, and graphics cutovers. As mentioned earlier in the discussion around the cutover database fields, it is critical that the software, interlock, and wiring relationships between points, loops, logic, and graphics be clearly understood. For example, if the same signal is both an input to a loop and to complex logic, then your plan needs to include properly synchronizing or sequencing the cutover of both. It also needs to include cutover of other I/O points associated with the loop and logic in a timely manner. If you move the I/O point and the logic to the new system but haven't moved all of the other I/O points involved in the logic to the new system, your advanced logic will not function in either the old or new system.

Cutover plans are used to provide advanced communication of daily cutover details to the operations and cutover team. These plans are also the basis for directing the execution of cutover activities. If properly designed they can also be used as the foundation for cutover progress reports, which will be discussed later in this chapter.

PREPARED FIELD TEAM

The experience and preparation of the field cutover team make significant contributions to cutover success. The field cutover team's efficiency will allow for a faster cutover pace and their attention to detail will prevent wiring mistakes and other errors that can derail a cutover. Field resources must be comfortable with the loop checkout process and have a thorough understanding of how they will drive or measure signals of various I/O types.

Preparation should be made by the team in advance of each cutover day, which can be difficult given the extended nature of most cutover execution days. One way to address this challenge and adequately handle cutover preparations is to designate two field cutover leads that alternate days. If Team Leader A is working the cutover execution on a given day, Team Leader B is becoming familiar with the I/O being migrated the next day. Familiarity includes knowing which points will be cutover, where physical instrument and associated junction box locations are in the field, and what marshalling panels and I/O cabinets will be involved. Team leaders should also establish communications with the field operator of any impacted areas discussing and identifying any operational tasks that might need to be done as part of the field cutover activity.

The field cutover team should also have the knowledge and ability to quickly troubleshoot and resolve instrument, signal, power, and grounding issues related to the cutover. Preparation by the field cutover team can speed cutover, minimize issues, and reduce troubleshooting time when issues do occur.

CONTROL ROOM LEADERSHIP

Strong control room leadership for cutover activities is critical. The individual or individuals staffing the control room during cutovers and checkout often have a number of important responsibilities. The lead control room resource typically manages the daily cutover execution process with the authority to adjust the daily cutover plan as needed. For example, a particular pressure control loop was scheduled for cutover but in conversations with operations the control room leader discovers that operations is in the middle of a vessel flush that relies on this pressure control loop. As a result, the control room lead postpones the cutover of that loop until later in the week.

The cutover control room leader is also responsible for coordinating cutover team activities with a designated operations representative and initiating the cutover of a particular I/O point. The control room leader engages the field construction team and helps direct activities. He is also responsible for verifying the control system functionality as part of the cutover activities.

In most cases, the cutover control room leader is the control system engineer who configured the system. This is beneficial because he can help troubleshoot any control system configuration issues found during the checkout process. There may be supporting resources that the control room cutover leader also utilizes. For instance, an engineer or operator may be stationed at the old operator console to record the process value and verify that the signal becomes deactivated without impacting any other functions in the old control system.

The control team leader may also be tasked with documenting daily cutover progress. The project manager or another designee may handle this function on a formal basis, but the control team leader's input is required for effective progress tracking. It is helpful for the control room cutover leader to be comfortable with the old control system, new control system, and the process operations or have resources assisting who have this knowledge.

The reporting function can be an extension of the existing cutover plan spreadsheets when advanced thought is given to the design. This simplifies the reporting process and makes it easy for the control team leader to handle the daily progress reporting function. Figure 9.3 below provides a sample daily cutover progress report reflecting one approach that can allow for simple tracking and reporting.

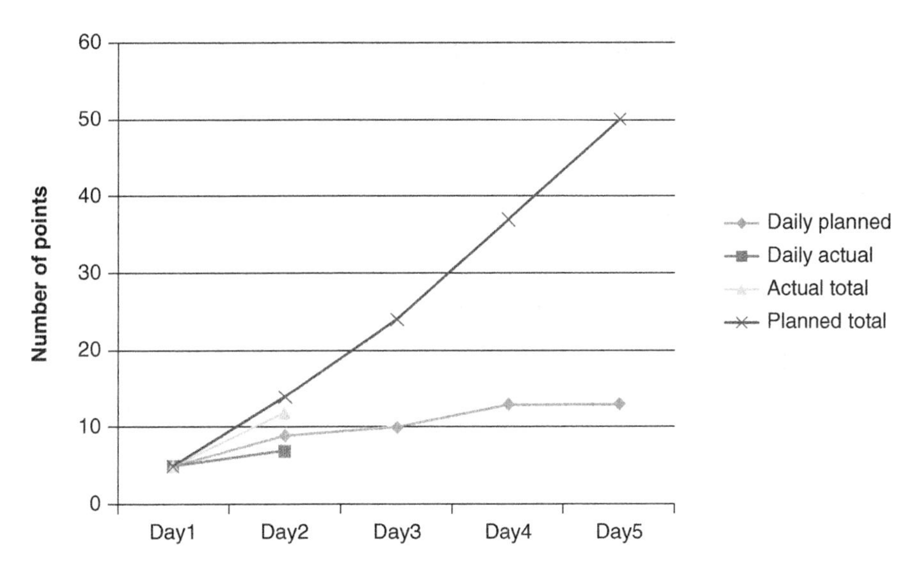

Number	Tag	Description	Point type	Status	Notes
1	Fi-100	Solvent flow to RX1	Analog input	Complete	
2	Ai-102	RX1 off-gas analysis	Analog input	Complete	
3	Pi-100	RX1 pressure indication	Analog input	Complete	Control loop pic-100 input
4	Py-101	RX1 pressure control valve	Analog output	Complete	Control loop pic-100 input
5	Ti-100	RX1 solvent temperature	Analog input	Incomplete	Not displaying properly on RX1 graphic
6	Li-100	RX1 level indiciation	Analog input	Complete	
7	Hs-104	Catalyst pump handswitch	Digital output	Incomplete	New catalyst batch in progress, postponed
8	Lah-100	RX1 high level alarm	Digital input	Complete	Used in RX1 interlock logic
9	Ti-101	RX1 cooling jacket temp.	Analog input	Complete	

Figure 9.3. Sample daily cutover progress report.

On a single graph this reporting format tracks the number of points planned for cutover on a given day, the actual points cutover on a given day, the overall sum planned for cutover, and the total sum of points actually cutover. This allows you to quickly compare the actual versus planned on both a daily and overall basis. The daily cutover spreadsheet below the graph identifies the specific I/O points planned for a given day of cutover and the associated status.

The example case outlined in Figure 9.3 is a cutover, which includes 50 total I/O points. The plan for Day 1 was to cutover five points, which was accomplished. On Day 2, the current day, the plan was to convert nine points, but only seven were converted. This lag is now reflected in the graph both through the daily actual versus planned and the total actual versus planned overall tracking.

STRONG OPERATIONS COORDINATION

The awareness and involvement of operations with cutover activities is a shared responsibility of the project team and operations management. Operations management has the responsibility to work with the project team to provide a staffing plan that supports cutover activities. As a given point is moved to the new system, operations must have the control room staffed to support the transition of process control and monitoring from the old system to the new system.

It is recommended that operations designate an individual to coordinate with the project team in the control room during the cutover. This may be the on-duty shift supervisor or it may be a single designee across all shifts such as a senior operator. This individual is given the daily cutover plan and reviews it for potential conflicts with production activities prior to starting cutover activities. The operations coordinator also handles securing the necessary permitting for cutover activities each day. Finally, this operations representative is notified as each point is cutover and ensures that information is communicated to the appropriate operational resources in a timely manner.

Field operations staffing is also required to support cutover activities. The field operator in a given area assists the cutover team in finding physical instrument locations as part of cutover preparation. Field operators may also be needed to perform operational functions as part of the cutover process. For example, a hand-off-auto switch may need to be positioned in the hand or off position briefly while an associated point is converted to the new control system. This must be done by operations rather than the field cutover team.

The preferred staffing approach for field operations is to have one or more dedicated field operating technicians assigned to support the cutover project. They are invaluable resources who simplify and speed up both field cutover team preparations and cutover execution activities. However, extra operations resources are often not available either because of lean staffing or cost limitations. In these cases, the cutover activities need to be coordinated with the on-duty field operator for a particular area so as not to conflict with their ability to perform their other required job functions.

While operations management is responsible for providing staffing to support the field and control room cutover activities, the project manager or a designee such as the control room cutover leader must correspondingly initiate communication, collaboration, and coordination with operations. The control room team leader provides daily cutover plans in advance to operations so that they can be reviewed and approved. Each day, a cutover update or progress report should be available in the control room as well. As cutover activities are initiated for I/O points the control room cutover team leader must actively communicate with the operations representative or respective control room operator. Good coordination and frequent communications are integral to seamlessly transferring points from the old system to the new system without production incidents.

COMPLETE LOOP PACKAGES

Complete loop packages help facilitate efficient and incident-free cutovers. These loop packages should provide all of the details associated with a loop, which in this case not only includes control loops but each I/O signal to the control system. This enables the field cutover team to quickly reference and review the current loop information associated with the existing control system, all changes to that loop for the new control system and what steps should be taken for the cutover.

Thorough loop packages should include the following:

- Old loop sheet for the existing control system
- New loop sheet for the new control system
- Notes for any detailed instructions or special considerations
- Control Loop or I/O point checkout form.

To reduce paperwork, some companies compile all loops or points cutover in a day into a single, overall loop checkout form rather than having individual forms. While this reduces paperwork, it can circumvent the intent of the form, which is to promote a step-by-step checkout process and create a documented record of the checkout. For this reason, I recommend keeping individual loop or point checkout form documents as part of the loop package.

Loop packages should be created for each individual loop or point and grouped in folders for the cutover team according to cutover plans by day, area, or another logical factor. It is essential that the field cutover team redline any changes to the loop documentation when deviations are identified or changes required. Documentation of these changes will need to be included in final project documentation preparation and can easily be overlooked if they are not immediately captured.

At the end of each day, the loop folders for all points that have been cutover should be submitted back to the project manager, control room cutover lead, or a specified designee. They should be kept in a centralized location so that they can be referenced if a problem subsequently arises with the loop during the cutover period. Once the cutover period is complete and the unit is operating on the new control system, all loop folders are sent to the instrument, electrical, and controls design teams who review them for any changes and update the final project loop drawings and documentation.

EFFICIENT CHECKOUT PROCESS

The process of checkout and verification of functionality for I/O points as they are converted to the new system is a critical quality assurance step. On hot cutovers this is done on a per point basis with active operations so it can easily be determined whether a signal is accurate. For example, if you cutover a temperature indicator and you recorded a process value of 200°F on the old system before cutover then if the signal is reading 200°F on the new system when it is connected it is likely working. This assumes no major process operational shifts or control changes during the brief time between removing the point from the old system and connecting it to the new system, which is why communications with the operator is so important. You may still want to perform additional range checks, but by recording the value before disconnect and comparing the value after reconnect, you get a quick validation of a working signal.

For hot cutovers, a reasonable rule of thumb is conversion of forty I/O points per day. This would include a mixture of control and indication points, which is why I prefer using I/O points as the measure rather than control loops. I have been on projects where we have far exceeded these numbers and on others where we lagged behind these averages. The daily cutover rates are influenced by the experience of personnel staffing the cutover, the complexity of the loops or logic, and the nature of the process operations.

Ultimately, the cutover pace is determined by operations. They must be able to manage process operations from the new interface for points that are cutover. If you can cutover 90 points per day, but it is overwhelming to the operator responsible for monitoring and controlling from the new system, then your pace will need to be scaled back to a point where it is practical for the operator to handle adapting to the new system.

There are many different variations of I/O point and control loop checkout forms. They provide a step-by-step instruction for a checkout and contain a signoff for each step. A simplified form of a completed analog input checkout form lacking final signatures is shown in Figure 9.4 below.

There are numerous loop checkout processes that are employed based on the signal type, cutover methodology, and the end user company checkout philosophy. For example, for a cold cutover on analog input signals, you may want to drive a signal from field device or a field junction box near the device. You would then validate that the analog signal at the operator workstation for the new control system displays the expected value. This is commonly done over a distributed signal range to ensure that the device is ranged properly and setup appropriately as a linear or square root signal to function properly across the range.

Performing a checkout of logic as part of the cutover process can be complex but also serves as an important functional verification. While software

Analog Input Checkout Form

Instrument Tag: FI-100 **Date: 3-12-13**

Step	Details	Initial	Notes
1	Instrument manufacturer, model and tag matches loop documentation and instrument specification	DAR	
2	Instrument configuration: Square Root ______ Linear___X___ Range: _0_to _30_ in H2O Engineering Units: GPM Engineering Range: 0 to 300 GPM	DAR	
3	Electrical and Grounding Verified	DAR	
4	Labeling Verified at: Instrument ____X____ Junction Box ___X___ Marshalling Cabinet ___X___ I/O Cabinet ___X____	DAR	Cabinet Label at Marshalling Panel was missing.
5	Terminations Verified at: Instrument ___X____ Junction Box ___X____ Marshalling Cabinet ___X___ I/O Cabinet ___X____	DAR	
6	Signal Simulation From Field Device: Simulated Value DCS Reading 0% (4 mA) 0.1 GPM 50% (12 mA) 150.2 GPM 100% (20 mA) 300 GPM 0% (4 mA) 0.1 GPM	DAR	Verified at control system via faceplate and on Utilities graphic
7	Red-lines in loop folder? Yes ___ No ___	DAR	Red-lines on loop drawing as I/O card addressing was incorrect
8	Other Comments:		
I&E Technician Signoff: ____________ Date: ________ Supervisor Approval: ____________ Date: ________			

Figure 9.4. Example analog input checkout form.

simulation is sometimes used to verify logic functionality in testing prior to the cutover, a checkout with live device signals should be done whenever possible as well. It is also vital to complete a thorough checkout of graphics. It is convenient to call up individual point faceplates for the loop checkout process because they have all of the information for a given point summarized in a single location. However, each graphic that displays a point or has a dynamic function that triggers from a point parameter also needs to be checked to make sure that the information is properly displayed. Engaging an operator to assist with these activities can be a great training opportunity and will help them gain experience navigating in the new system.

The loop checkout process defines clear step-by-step instructions for verifying a point is working in the new control system. It also serves as the final quality assurance step before a point is put into service in the new control system. Finally, it documents and establishes a historical record of the checkout. Depending upon your industry and global location, regulatory agencies such as the Occupational Safety and Health Administration (OSHA), Pipeline and Hazardous Materials Safety Administration (PHMSA), or the Food and Drug Administration (FDA) might require these records be available during audits.

SUMMARY

The cutover is the final major step in a control system migration project. A poorly executed cutover phase is capable of causing a control system migration project to fail even when the design, engineering, and construction of the project leading up to the cutover are successful. The cutover should be considered a culmination of all aspects of the control system migration and managed as a small project within the overall project. Successful cutovers fit within the context of the overall migration projects scope, schedule, and budget requirements and minimize disruptions to process operations.

The decision of whether to use a hot, cold, or blended cutover methodology helps shape how your overall project is executed. In some cases, the cutover methodology is determined based on a significant overriding factor such as the critical nature of the process operations. Cutovers require a tremendous amount of planning and detailed design before being executed. Physical verifications during the design phase can help to avoid many of the common issues which arise during cutover.

Cutover execution success is largely based on the field team and control room cutover leader working effectively together while proactively coordinating and communicating with operations. A prepared field cutover team with complete loop packages and a clear understanding of I/O checkout processes can speed the rate of cutover and avoid wiring mistakes. Just as essential is strong leadership in the control room throughout the cutover. The control room cutover leader communicates with operations, handles control system checkout, coordinates the checkout process, and often manages the tracking and reporting of daily cutover progress.

Three Key Takeaways

- Determining the best cutover methodology for your migration project involves many considerations, including the nature and complexity of the process operations, the I/O point count, logic and configuration complexity, cutover cost, outage constraints, and the physical control room environment.
- The cutover plan is a critical document, which should address the work to be done, which resources will be doing what portions of the work, when the work is planned and how the work is sequenced and executed.
- The preparation, teamwork, and efficiency of the field cutover team and the control room leadership combined with strong operational coordination are significant factors in determining cutover success.

10

Project Closeout and Lifecycle Management

When your control system cutover is complete, the system is transferred to operations. The project team may still be involved with the completing a small number of remaining scope items or making configuration changes that are requested during early operations with the new control system. Typically, your migration project enters the closeout phase around this time.

Cutover is a major milestone in any control system migration project and it is natural to relax when it is complete. While it may seem like the project can easily be managed post-cutover, it is not uncommon to see costs trend over budget during the closeout stage of the project. In many cases such as with EPC firms, vendors, and system integrators, their project staffs are required to be billable. When staff is transitioning off of your project and ramping up on another project, time often gets charged against your project as a fallback with limited productivity for these billable charges. It is important to aggressively manage billing charges during the project closeout phase of your project.

Another common reason for overages during the closeout period is scope expansion. The back end of the project is an opportune time for additional scope items to creep into your project. Many of the operational support items that arise during the first few weeks of running the process unit with the new control system are not clearly defined. As a result, there is a tendency to see the scope shift from support items to added functionality or enhancements. The project staff may have some free time available, so they are more receptive to handling items that are out of scope without proper change order documentation. This can burden your project with unforeseen and offplan costs. More importantly, if the changes are substantial, it can actually circumvent the end-user company's management of change process.

Project closeout activities also frequently drag out, extending the schedule and delaying formal project completion. When schedule durations are extended, productivity drops and costs increase. These delays often occur because the proper closeout staffing plan is not in place. For instance, many times one designer is responsible for completing drawing updates and assembling the final project documentation, while the remainder of the design team is assigned to other projects. Maintaining sufficient project design staff to quickly complete the final project documentation is often a better approach.

To avoid these common late-stage project issues, it is important to define a project closeout process that establishes clear milestones, prevents unnecessary labor charges, and eliminates indefinite ongoing work items. The project closeout process represents a formal delivery and acceptance of the project signaling the completion of the intended scope. If the project utilized one or more third-party providers for the EPC scope, then establishing completion of the defined project scope is required from a contractual obligation standpoint and allows for final invoicing.

Project closeout activities are not isolated to the end of the project but instead can take place throughout a project. When an individual milestone task is completed, you can have a formal review and acceptance of that work and the associated deliverables. This approach is beneficial because it helps validate earned value analysis, provides a good means for checking quality, and documents acceptance throughout the project simplifying final closeout. You might also choose to close billing codes associated with individual milestones so that no additional charges are made against that particular work breakdown task, in effect providing a financial closeout of the milestone.

The key elements of a project closeout process are shown in Figure 10.1 below.

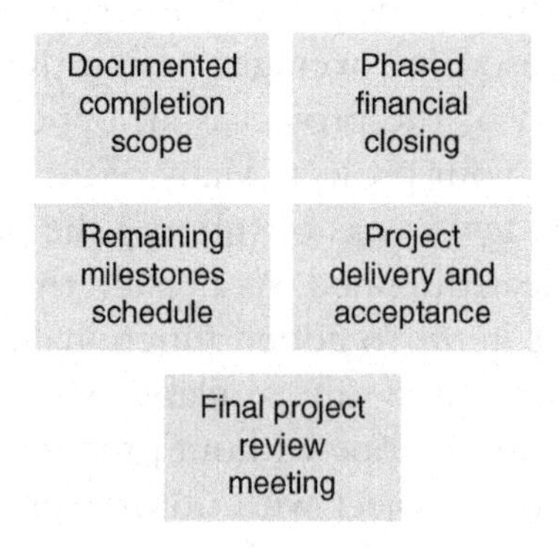

Figure 10.1. Essential elements of the project closeout process.

In the remainder of this chapter, we will discuss these essential project closeout process elements and the reasons they are important. We will review some of the common challenges related to the project closeout process. Finally, we examine how to transition from your migration project to successful long-term operations, care, and maintenance of your new control system using a lifecycle management program.

DOCUMENTED COMPLETION SCOPE

When you reach the closeout phase of your migration project, the remaining scope should be limited and typically includes the need to:

- Make any control system configuration changes or adjustments needed
- Complete any remaining demolition activities
- Finalize and update project documentation

While there can be other miscellaneous items that might be involved with your final scope, if it is anything substantial outside of these areas you might need to examine whether you are actually in the closeout phase. Because the scope at this stage should be minimal, the remaining work activities should be converted to a punch list format as shown in Table 10.1 below.

Table 10.1. Project closeout scope punch list

Item	Task	Status	Date Completed
1	Change text formatting on utilities area graphic to match others	Complete	April 1, 2013
2	Remove old marshalling cabinet 1A-5 and move to warehouse	Complete	April 16, 2013
3	Update loop sheet drawings per cutover field team red-line markups	Pending	
4	Assemble and submit final project documentation package	Pending	

Using this punch list approach consolidates the remaining scope into a single location clearly identifying the specific steps required to complete the project and making it easier to track progress. After the punch list is created and agreed to by all parties, no scope items should be added to the list. This will help eliminate additional scope creep at this final project stage. The action items in the punch list should also have budgets and schedules attached to them like any project tasks.

Defining the project closeout scope items associated with control system configuration changes can be especially nebulous as demonstrated in the example below.

Example

You have completed cutover and your migration project is winding down. The control system engineer on your project is providing support to operations and making necessary configuration changes. The operations manager makes a request that control system engineer modify all of the pump symbols and associated dynamic status colors because the configured symbols are confusing to the operators. The engineer notifies you and estimates the changes will take approximately two days because the pump elements will need to be customized, applied to the graphics and checked. You must determine whether this is part of the existing scope or should be handled as a change order.

This presents a real dilemma between balancing the project budget and scope with the long-term operational needs. You might be justified in approaching these modifications as additional scope requiring a change order. After all, an operations representative signed off on the graphics earlier in the development process. However, you also want to ensure that the configuration of the new control systems meets operational requirements. You especially want to avoid any production incidents related to either control functionality or how information is presented in the new control system. In the end, you determine it is in the best interest of the project to accommodate the changes within the existing project scope.

Closeout configuration issues like the one presented in the example above occur with some frequency. It is natural that as you gain experience operating a system, you will discover adjustments that will make it easier to operate. You should ensure that changes during this process do not conflict with the design basis or violate company management of change processes. Most migration projects are performed under an umbrella management of change form for the project duration. However, if changes during the first few weeks of operation with the new control system are significant rather than minor adjustments, then they might require a standalone management of change form.

It is difficult and time-consuming to manage operational changes on a per item basis evaluating each against the existing scope. The best way to avoid this challenge is by defining the support scope in terms of manpower and time rather than specific detailed configuration tasks. For instance, you might agree to include one controls engineer for two weeks and one graphics designer for three days to support initial operations with the new control system. This provides you with a defined budget burden for the project and should be more than enough time to address loose ends and make any configuration tweaks.

PHASED FINANCIAL CLOSING

One of the most effective ways to keep on budget throughout your project is by using a phased financial closing approach. Some project financial accounting systems allow this to be easily done, while in others it can be more difficult. This methodology requires a reasonably granular work breakdown structure with specific budgets and associated charge codes allocated to lead tasks. This is generally necessary to help you track and control project spending anyway.

Phased financial closing means that when a given work task is complete, you close the associated billing code so that no additional charges can be incurred against that task. A work task should not be considered complete until the work quality has been verified and it has been accepted by the appropriate individual within the end user company. This is essential because it can create accounting difficulties and add complexity if you need to make changes on a work task deliverable after you have closed the billing code option. Employing this approach throughout your project means that when you reach the closeout phase you have a minimum number of billing codes open. This makes it much easier to identify any unexpected cost burdening the project during closeout activities.

The budget for closeout activities is often a composite of design, construction, and controls engineering tasks. While it can be managed using the existing budget structure, it can be helpful to close all existing charge codes for the project and open a closeout budget as shown in Figure 10.2 below.

Billing Code	Project Tasks	Budget	Spent	Line Item % Spending	Status
1001	Configure I/O Points	$10,000	$9,000	90%	Complete Closed
1004	Configure Graphics	$20,000	$19,000	95%	Incomplete Closed
2005	Build Loop Sheets	$15,000	$14,000	93%	Incomplete Closed
3006	Cabinet and Workstation Demolition	$7,500	$5,000	67%	Incomplete Closed
5000	Closeout Punchlist Items	$5,500	$0	0%	Open

Open New Bill Code for All Punch List Items

Close Individual Tasks and Include Remaining Work Under New Billing Code for Closeout Punch List Items

Figure 10.2. Example closeout budget.

In this example, even though some of the work task may be incomplete, the individual line item work tasks are closed and the work is included under a newly opened task that covers all remaining closeout punch list items. The budget can be estimated based on the closeout work tasks remaining or it can be an accumulation of the remaining funds from the individual line items.

The creation of a separate closeout budget is generally only necessary if you have numerous resources completing closeout activities or expect the schedule to be a significant duration. The closeout budget is established based on the closeout punch list activities. All costs for the closeout phase are then funneled to a single charge code, which can simplify budget tracking at this late project stage.

REMAINING MILESTONES SCHEDULE

When project closeout action items are established, they should be given target completion dates. They should also be defined in terms of duration with any sequencing or timing requirements detailed. For example, changes to the final I/O index for the project might need to be made prior to completing updates on the loop sheet drawings.

When staff resources are pulled from the project after cutover, it can be easy to lose momentum. Insufficient project staffing can result in project completion delays. Once you have your punch list and have assigned a schedule to the action items, I recommend that you evaluate whether it is a reasonable time to completion. As a rule of thumb, project closeout activities as defined here should add 5–15% to your overall project duration. For instance, if your project is 40 weeks in duration, then you should be able to complete all of the project closeout punch list items in 2–6 weeks. Longer durations for project closeout usually indicate that you are understaffed in the closeout phase, or that your project had poor design quality, required significant field changes or the system configuration was inadequate resulting in a large amount of closeout work.

PROJECT DELIVERY AND ACCEPTANCE

While operations is using the new control system to operate the unit, it does not necessarily mean that they have formally accepted ownership of the system. It is not uncommon to see operations relying heavily on project team support for a period of time after cutover. However, at some point a clearly defined handover must take place and the project team's responsibilities must be complete. This handover transitions responsibility for the engineering, operation, and maintenance of the new control system to the appropriate departments within the end user organization. At this point, any additional changes related to the control system are handled through normal company management of change processes.

A significant part of the handover process is completion and delivery of the final project documentation. The materials included and the way they are delivered varies widely depending on specific company requirements. For example, many companies only require electronic delivery of project documentation today as opposed to hardcopy manuals. Table 10.2 below list is a checklist of common materials included in final project documentation packages.

Table 10.2. Final project documentation checklist

Description	Included (Yes/No)
Purchase orders and contracts with all vendors and service providers	
Scope of work	
Project schedule (final)	
Budget spreadsheet	
Fat procedures and results	
Sat procedures and results	
Project progress reports	
Training plans with all required documentation of attendance, testing, etc.	
Key correspondence or communications documenting project decisions	
Project I/O index	
Project drawings (new, revised, demo)	
Project drawings (temporary)	
Cutover plan and daily progress reports	
Cutover loop (I/O) folders with signed loop check forms	
All task completion signoff forms (e.g., Graphics)	
All equipment installation and maintenance manuals	

To complete project acceptance, a formal signoff form is recommended, which clearly states that the project team has met its responsibilities. Figure 10.3 below shows a simple example project acceptance form.

This form is especially critical if third-party service providers managed and executed the project as it documents completion of the contract. The form does not need to be lengthy or complex. The existing project documents such as the scope of work or change orders can be used as supplementary reference attachments to help define the project.

Final Project Delivery and Acceptance Form

I acknowledge completion of all work activities included as part of the defined scope of work for Project 1234 also referenced as the Control System Migration Project. I also acknowledge that all work was performed to an acceptable level of quality.

Individual Work Components

The following work components were part of the scope and have been evaluated and accepted:

- ☐ Instrumentation Design and Engineering
- ☐ Electrical Design and Engineering
- ☐ Controls System Design and Engineering
- ☐ Civil Engineering
- ☐ Field Construction

Attachments:

Attachment A: Scope of Work Document 1234A
Attachment B: Change Order 1234CO-1
Attachment C: Change Order 1234CO-2

Authorizations and Approvals

Name: Joe Smith	Name: Steve Brown
Title: Project Manager	Title: Operations Mgr.
Signature: Joe Smith	Signature: Steve Brown
Date: April 25, 2013	Date: April 25, 2013

Page 1 of 1

Figure 10.3. Simple example project acceptance form.

FINAL PROJECT REVIEW MEETING

The last active project step is to hold a final project review meeting with all key stakeholders and project team members. The final review meeting should serve to clarify and resolve any outstanding project-related items whether commercial, technical, operational, or otherwise. At this late stage there should not be many, if any unresolved issues. A sample final review meeting agenda is outlined in Figure 10.4 below.

The project manager should provide the team with a complete analysis of the project performance measuring execution against plan for the scope, schedule, and budget. The amount of information shared depends on how the project was executed, who executed the project, and the type of contract in place. For example, if it was a firm price project completed by a system integrator, they might not share a financial budget analysis of the project.

The design and engineering team as well as the project manager should lead a review of the final project documentation for completeness and to

Final project review meeting agenda
week of April 25, 2013

Resource	Duration	Subject
Project manager	15 minutes	Project performance metric review versus scope, schedule and budget
All	10 minutes	Identify, discuss and resolve any open items
Lead I&E engineer, lead controls engineer, project manager	25 minutes	Summarize and review final project documentation
All	20 minutes	Brainstorm project strengths
All	20 minutes	Brainstorm project opportunities for improvement
Project manager	5 minutes	Closing comments

Figure 10.4. Sample final review meeting agenda.

clearly communicate to the appropriate parties where to find information and resources related to the project and the new control system. Finally, the entire team should do an assessment of the project identifying what was done well and what could have been improved. It is essential to use each project as a learning opportunity for future improvements.

LIFECYCLE MANAGEMENT

The selection and installation of your new control system create the prime launching point for your lifecycle management program. While a lifecycle management program is likely not a formal part of your migration project scope, it is important for maintaining an ongoing positive perception of the project and relevant to the continuing successful operation of the new control system. Your migration project involves or impacts lifecycle management in several areas. First, as mentioned in an earlier chapter, the decision process for selecting your new control system should take into account the lifecycle status. For example, a system that has been on the market for 15 years might not be ideal if you believe that it is on the trailing end of its lifecycle and would require additional investment in upgrades or changes in the near future.

In addition, the completeness and accuracy of the control system documentation generated during the migration project are an essential part of establishing your lifecycle management program. Your new control system vendor likely provides a service and support program of some type. Vendors have a variety of definitions for control system lifecycle management. From the perspective of the end user company, it can best be defined as the ongoing and proactive service, support, upgrade, and parts replacement strategies, which balance total cost of ownership, return-on-investment, and optimized control system performance to meet key business goals. The lifecycle management program is applied to both software and hardware components of the process control system at all system levels such as workstations, controller, I/O, and networking. End user companies must decide how to apply their vendors offering to bring the most value to them.

I encourage end user companies to create a control system lifecycle management plan with involvement from the project team prior to the completion and closeout of the migration project. This will leverage the expertise and knowledge of the project team with the details of both your control system and your specific applications. The control system vendor should also be involved in the plan development so that they can better understand your needs and also offer suggestions and insights based on their lifecycle services programs and experiences with other clients. Your lifecycle management plan should consolidate information about the new control system into a single document and provide recommendations on system care and maintenance. Some key elements to consider for your initial lifecycle management plan are listed and described in Table 10.3 below.

Table 10.3. Initial lifecycle management plan elements

Element	Description
Installed hardware inventory	A list of all hardware installed with model and version numbers
Installed software inventory	A list of all software installed including version and patch numbers
Spare parts inventory and philosophy	A list of all required system spare parts and an explanation of how to determine what spares are kept and in what quantities
Software update and testing philosophy	An explanation of how software updates will be applied and tested (e.g., Only major software releases, updates every year or case-by-case evaluations?)
Hardware update and testing philosophy	An explanation of how hardware updates will be applied and tested
Hardware obsolescence and phase out strategy	An explanation of how hardware which the vendor places on an obsolescence path will be phased out. For example, will you immediately replace, will you purchase spares so you can leave in place, or will you take some alternate strategy?

(*Continued*)

Table 10.3. (*Continued*)

Element	Description
Engineering services philosophy	A description of the preferred strategy for making system configuration changes including what resources are responsible
Maintenance philosophy	A description of the preferred frequency, timing, planning, and execution strategies of regular control system maintenance activities
Technical support plan	An explanation of how staff resources get technical questions about the control system answered. For example, you may have a technical support agreement with the vendor.
New technology evaluation and adoption process	Outlines the process to be used to evaluate and make decisions about the adoption of new technologies. For example, if the vendor releases the next generation of operator workstation do you migrate to that version?

Strong lifecycle management programs can extend the life of your new control system while enhancing system performance and reliability. Initiating an effective lifecycle management program at the end of your migration project will enable you to continually optimize the operational value of your new control system. Lifecycle management is a multifaceted, complex program. We summarize it here particularly as a natural extension of a migration project. However, I encourage you to read additional and more in-depth materials on lifecycle management planning before establishing a program.

Control systems with weak lifecycle management systems are often characterized by one or several of the following attributes:

- Disparate versions of software and hardware
- Higher total cost of ownership
- Poor reliability, high system inefficiency
- Reduced performance
- Shorter life spans

You can quickly find your new control system in a poor lifecycle management position and the primary culprit is most frequently a lack of ongoing investment. End user companies must recognize that control systems involve recurring financial investments and proactively managing their well-being is essential to their continued effectiveness. This can often be a challenge sell within an end user organization as described in the example below.

Example

You recently completed a control system migration project installing a new control system at your facility. You did not create a lifecycle management program at the end of the project nor have you yet established a maintenance support agreement with the control system vendor as all parts are under warranty. Three months after your project was complete, the control system vendor issued a major release of software for the operator workstation. This software improves the maintenance diagnostics and enables better alarm tracking. A firmware release for the controller was issued six months after your installation.

You consider several options including leaving the system at its current software version and postponing any updates. If you decide to immediately implement the updates you will either pay for the vendor to provide the services, sign up for a maintenance agreement, or implement the upgrade yourself. Recognizing the benefits of some of these upgrades, you approach the operations manager to get approval for a purchase order to implement the upgrades. The operations manager denies your request because of the large investment already made in the system.

In the example above, the operations manager likely has budget limitations or other priority spending items. It is natural to look at the large investment just made in the control system and question either the need for or timing of additional investment. This is one reason it is important to discuss what it takes to support a new control system over the lifecycle as part of the migration project.

Vendors are continuously working to improve their control systems. It is up to each individual control system owner to selectively choose which technology enhancements, system upgrades, and software updates to apply. However, by establishing a documented lifecycle plan and proactive approach to managing updates, you have guidance for your decision process. With the faster pace of technological changes today, it is more important than ever to have a lifecycle management program in place to both prevent component obsolescence issues and take advantage of enhanced functionality. The return-on-investment of your new control system can quickly deteriorate if you are not continually evaluating the status of your system.

SUMMARY

Project closeout is a process by which formal delivery and acceptance of your control system migration project takes place. The process includes:

- Documenting and completing the remaining scope
- Phasing the financial closeout of the project

- Establishing final milestone completion schedules
- Delivering final project documentation and gaining formal acceptance
- Holding a final project review meeting

Project closeout activities can often begin in the early stages of your project when you complete an individual milestone. Formal review and approval of individual task deliverables provide a means for quality checking and also helps validate earned value analysis. Individual milestone completions are also an ideal time to close work breakdown tasks to additional billing, which helps manage and control spending.

The completion of your control system migration project is the natural starting point for initiating a lifecycle management program. Control system lifecycle management programs with the high frequency of technology enhancements and changes today are more essential than ever. The purpose of good lifecycle management programs is to increase the longevity and effectiveness of your control system while optimizing your total cost of ownership. While your migration project has provided you with a new control system, your long-term return-on-investment will be determined by how you operate and maintain the system over its life.

Three Key Takeaways

- The project closeout process formalizes delivery and acceptance of the project signaling the completion of the intended scope.
- Utilizing a phased financial project closeout where you close billing associated with already completed project tasks can help you control spending on your project.
- Your migration project is the ideal point to begin a control system lifecycle management program, which is the ongoing and proactive service, support, upgrade, and parts replacement strategies which balance total cost of ownership, return-on-investment, and optimized control system performance to meet key business goals.

Supplemental Resource List

Below is a list of supplemental resources including magazine and industry association websites as well as articles providing information focused on or in some way related to control system migrations. This list is by no means all-inclusive and should not be considered an endorsement for any of the listed sites. Third-parties provide and control the content of the web links which are subject to change. Therefore, the author and publisher accept no responsibility and make no guarantees as to the accuracy of content listed below.

WEBSITES

www.arcweb.com
www.automation.com
www.automationworld.com
www.chemicalprocessing.com
www.controleng.com
www.controlglobal.com
www.controlsys.org
http://www.ieeecss.org/
www.isa.org

ARTICLES

Bill Lydon. "Automation Life Cycles-Important Consideration for Purchases, Migrations & Upgrades" (March 29, 2013). http://www.automation.com/automation-news/article/automation-life-cycles-important-consideration-for-purchases-migrations-upgrades

Bill Lydon. "Automation Upgrade and Migration Investment Strategies." (July 25, 2011). http://www.automation.com/automation-news/article/automation-upgrade-and-migration-investment-strategies

Bill Lydon. "DCS Migration & Retrofit Insights." (April 17, 2012). http://www.automation.com/automation-news/article/dcs-migration-retrofit-insights

Bill Lydon. "System Migration Attacks Skills Crisis." (April 20, 2010). http://www.automation.com/library/articles-white-papers/articles-by-bill-lydon/system-migration-attacks-skills-crisis

Christopher A. DaCosta, and Ken Keiser. "What cost migration?" *InTech.* (September 2007). http://www.isa.org/InTechTemplate.cfm?Section=Communities2&template=/TaggedPage/DetailDisplay.cfm&ContentID=63838

Dan Hebert, PE. "Best practices in control system migration." *Control.* (January 2007). XX no. 1.

Dan Roessler, and Lisa Garrison. "Five practical elements of effective SCADA graphics." *Pipeline and Gas Journal* (February 2013), 240 no. 2. http://www.pgjonline.com/five-practical-elements-effective-scada-graphics

Dave Woll. "The coming wave of process safety system migration." *InTech.* (May/June 2012). http://www.isa.org/InTechTemplate.cfm?Section=Communities&template=/ContentManagement/ContentDisplay.cfm&ContentID=89644

Eric Schnipke. "Hot cutover boosts control system migration."*Chemical Processing.* (March 2, 2008): 39–42. http://www.chemicalprocessing.com/articles/2008/067/

Grant Gerke. "Changing Workforce Drives Control Room Modernization." (January 28, 2013). http://www.automationworld.com/control/changing-workforce-drives-control-room-modernization

Ian Verhappen. "Migrate without a migraine."*Chemical Processing.* (March 8, 2012): 35–38. http://www.chemicalprocessing.com/articles/2012/migrate-without-a-migraine/

Larry O'Brien. "Modernization projects reduce field service cost." *InTech.* (September/October 2012). http://www.isa.org/InTechTemplate.cfm?Section=Communities&template=/ContentManagement/ContentDisplay.cfm&ContentID=90805

Marjorie Oschner and Walter Guy Wiles III. "A Path to Migration: Legacy Systems won't Last Forever; Planning Upfront is Key." (2008). http://www.isa.org/InTechTemplate.cfm?Section=Article_Index1&template=/ContentManagement/ContentDisplay.cfm&ContentID=71545

Mark Rosenzweig. "Make the most of control system migration." *Chemical Processing.* (July 2008): 7. http://www.chemicalprocessing.com/articles/2008/104/

Matt Sigmon. "DCS migration: Failure is not an option and doing nothing is not a solution." *Control* (December 2012), XXV no. 12: 41–42.

Mike Alsup. "Control systems: Don't get misled by modernization misconceptions." *Chemical Processing* (September 24, 2012): 38–46. http://www.chemicalprocessing.com/articles/2012/control-systems-don-t-get-misled-by-modernization-misconceptions/

Nigel James. "Reduce costs and risk by following these control system migration best practices." *Control* (January 2009), XXII no. 1: 63–66.

Paul Galeski, P.E. "Tidal Wave of DCS Replacement on the Horizon." (July 1, 2012). http://www.automationworld.com/control/tidal-wave-dcs-replacement-horizon

Peter Reynolds. "Suppliers and Users Share Responsibility for Successful Control-System Migrations." *Hydrocarbon Processing* (May 1, 2012). http://www.hydrocarbonprocessing.com/Article/3016872/Suppliers-and-users-share-responsibility-for-successful-control-system-migrations.html

Renee Robbins Bassett. "Operator Interfaces: Moving from Comfortable to Most Effective." (December 7, 2012). http://www.automationworld.com/operations/operator-interfaces-moving-comfortable-most-effective

Index

THIS TITLE IS FROM OUR MANUFACTURING ENGINEERING COLLECTION. OTHER TITLES OF INTEREST MIGHT BE...

Alarm Management for Process Control: A Best-Practice Guide for Design, Implementation, and Use of Industrial Alarm Systems
By Douglas H. Rothenberg

Advanced Regulatory Control: Applications and Techniques
By David W. Spitzer

Process Control Case Histories: An Insightful and Humorous Perspective from the Control Room
By Gregory K. McMillan

Protecting Industrial Control Systems from Electronic Threats
By Joseph Weiss

Industrial Resource Utilization and Productivity: Understanding the Linkages
By Anil Mital, PhD

THE WBF BOOK SERIES–ISA 88 Implementation Experiences, Applying ISA 88 In Discrete and Continuous Manufacturing, ISA 95 Implementation Experiences and ISA 88 and ISA 95 in the Life Science Industries

Robust Control System Networks: How to Achieve Reliable Control After Stuxnet
By Ralph Langner

Raw and Finished Materials: A Concise Guide to Properties and Applications
By Brian Dureu

Quality Recognition & Prediction: Smarter Pattern Technology with the Mahalanobis-Taguchi System
By Shoichi Teshima, Yoshiko Hasegawa, Kazuo Tatebayashi

Going the Distance: Solids Level Measurement with Radar
By Tim Little, Henry Vandelinde

Textile Processes: Quality Control and Design of Experiments
By Georgi Damyanov, Diana Germanova-Krasteva

Plant IT Integrating Information Technology into Automated Manufacturing
By Dennis L. Brandl, Donald E. Brandl

Mastering Lean Six Sigma: Advanced Black Belt Concepts
By Salman Taghizadegan

Automated Weighing Technology: Process Solutions
By Ralph Closs, Henry Vandelinde, Matt Morrissey

CPSIA information can be obtained
at www.ICGtesting.com
Printed in the USA
LVOW03*2329250816
501549LV00005B/7/P

9 781606 504437